TORQUES AND ATTITUDE SENSING IN EARTH SATELLITES

APPLIED MATHEMATICS AND MECHANICS

An International Series of Monographs

EDITORS

Volume 1. K. Oswatitsch: Gas Dynamics, *English version by G. Kuerti* (1956)

Volume 2. G. Birkhoff and E. H. Zarantonello: Jets, Wakes, and Cavities (1957)

Volume 3. R. von Mises: Theory of Compressible Fluid Flow, *Revised and completed by Hilda Geiringer and G. S. S. Ludford* (1958)

Volume 4. F. L. Alt: Electronic Digital Computers — Their Use in Science and Engineering (1958)

Volume 5. W. D. Hayes and R. F. Probstein: Hypersonic Flow Theory (1959)

Volume 6. L. M. Brekhovskikh: Waves in Layered Media, *Translated from the Russian by D. Lieberman* (1960)

Volume 7. S. Fred Singer (ed.): Torques and Attitude Sensing in Earth Satellites (1964)

Volume 8. M. Van Dyke: Perturbation Methods in Fluid Mechanics (in preparation)

TORQUES AND ATTITUDE SENSING IN EARTH SATELLITES

Edited by

S. FRED SINGER

NATIONAL WEATHER SATELLITE CENTER, WEATHER BUREAU,
DEPARTMENT OF COMMERCE, WASHINGTON, D.C.

1964

ACADEMIC PRESS · NEW YORK · LONDON

ACADEMIC PRESS INC.
111 Fifth Avenue, New York, New York 10003

United Kingdom Edition published by
ACADEMIC PRESS INC. (LONDON) LTD.
Berkeley Square House, London W.1

Library of Congress Catalog Card Number: 64-15274

PRINTED IN THE UNITED STATES OF AMERICA

List of Contributors

Numbers in parentheses indicate pages on which the authors' contributions begin.

J. ALISHOUSE, *U.S. Weather Bureau, Washington, D.C. (207)*

G. COLOMBO,[1] *Smithsonian Institution Astrophysical Observatory, Cambridge, Massachusetts (175)*

BARNEY J. CONRATH, *Goddard Space Flight Center, National Aeronautics and Space Administration, Greenbelt, Maryland (235)*

J. E. DELISLE, *Instrumentation Laboratory, Massachusetts Institute of Technology, Cambridge, Massachusetts (31)*

WILLIAM J. EVANS, *Grumman Aircraft Engineering Corporation, Bethpage, New York (83)*

ROBERT E. FISCHELL, *The Johns Hopkins University, Applied Physics Laboratory, Silver Spring, Maryland (13)*

E. HECHT, *Astro-Electronics Division, Radio Corporation of America, Princeton, New Jersey (127)*

B. M. HILDEBRANT, *Instrumentation Laboratory, Massachusetts Institute of Technology, Cambridge, Massachusetts (31)*

BARBARA KEGERREIS LUNDE, *Goddard Space Flight Center, National Aeronautics and Space Administration, Greenbelt, Maryland (221)*

R. A. LYTTLETON,[2] *Jet Propulsion Laboratory, Pasadena, California (107)*

W. P. MANGER, *Astro-Electronics Division, Radio Corporation of America, Princeton, New Jersey (127)*

ROBERT J. MCELVAIN,[3] *Space-General Corporation, El Monte, California (137)*

ROBERT J. NAUMANN, *Research Projects Division, George C. Marshall Space Flight Center, National Aeronautics and Space Administration, Huntsville, Alabama (191)*

[1] Present address: University of Padua, Italy.
[2] Present address: St. Johns College, Cambridge, England.
[3] Present address: Hughes Aircraft Company, Space Systems Division, El Segundo, California.

E. G. OGLETREE, *Instrumentation Laboratory, Massachusetts Institute of Technology, Cambridge, Massachusetts (31)*

GORDON S. REITER, *Dynamics Department, Space Technology Laboratories, Redondo Beach, California (1)*

ROBERT E. ROBERSON, *Department of Engineering, University of California, Los Angeles, California (73)*

S. FRED SINGER,[4] *National Weather Satellite Center, Weather Bureau, Department of Commerce, Washington, D.C. (99, 107)*

W. T. THOMSON, *Department of Engineering, University of California, Los Angeles, California (1)*

D. Q. WARK, *U.S. Weather Bureau, Washington, D.C. (207)*

DONALD D. WILLIAMS, *System Design Department, Project Syncom, Hughes Aircraft Company, Culver City, California (159)*

RAYMOND H. WILSON, JR.,[5] *NASA Goddard Space Flight Center, Greenbelt, Maryland (117)*

G. YAMAMOTO, *Tohoku University, Sendai, Japan (207)*

[4] Present address: School of Environmental and Planetary Sciences, University of Miami, Coral Gables, Florida.

[5] Present address: NASA Headquarters, Washington, D.C.

Preface

One of the most important, and at the same time, one of the most challenging problems in space technology is the proper orientation of space vehicles. For example, weather satellites and communication satellites function most effectively when the sensors or antennas can be oriented vertically downward. In the case of weather satellites, one would like to keep this vertical error to a small fraction of a degree of arc. On the other hand, astronomical observation satellites must have orientation accuracies which are much more precise. In considering how to achieve this aim, the first and most important question is a scientific one: What are the possible torques that can affect the angular momentum of an earth satellite? The second question to ask is: What natural phenomena exist for establishing a reference? The answers to both of these questions are intimately tied up with an understanding of the environment in which a satellite operates. This environment consists of the gravitational field of the earth, and possibly other bodies, the geomagnetic field, the remaining atmosphere, both neutral and ionized, and the solar radiation field. In addition, there are meteors, whose occasional impacts produce impulsive perturbing torques, and perhaps other effects that have not yet been clearly recognized.

The purpose of this volume is to pull together and elucidate at least the major effects that have been recognized to date, discuss their relative importance, and present a scientific basis, particularly to engineers, for the design of attitude control systems. It should be noted, however, that the discussion does not deal directly with the design of attitude control systems, but only with the external torques and with the means of attitude sensing.

We start with an important effect which is often completely glossed over in courses on classical mechanics. No real body is completely rigid; as it spins it will dissipate kinetic energy, owing to imperfections of elasticity. The effects of this dissipation result in a change of orientation, until the body ends up spinning about the axis of maximum moment of inertia. This effect can be made use of in various ways which are discussed in the article by Reiter and Thomson.

The next three papers deal with the torque effects of a gravitational field, and particularly with its gradient; a perfectly uniform field cannot produce a torque. The paper by Fischell discusses a particular method of gravity-gradient stabilization that uses a passive device, an ultraweak spring, to damp librations and thereby achieve vertical orientation. The paper by DeLisle, Ogletree, and Hildebrant, on the other hand, discusses the use of gyrostabilizers as semipassive devices to achieve internal dissipation and damping as well as certain desirable control over the orientation of the satellite. Finally, the paper by Roberson takes up the interesting problem of how to calculate the torque in the case of a generalized gravitational field.

The next group of papers deals with torques produced by the interaction of a satellite with its particle and radiation environment. The paper by Evans, for example, gives a technique for calculating the aerodynamic and radiation pressure torques for a satellite of complicated shape, while the paper by Singer deals with the Coulomb torque produced by the electrostatic "collisions" with ions and electrons in the ionized gas in which the satellite is moving. The paper by Lyttleton and Singer brings to bear on a specific example the question of internal energy dissipation and Coulomb drag.

The next set of papers discusses the effects of the geomagnetic field on the angular motion of a satellite. An introductory review is given by Wilson. The paper by Hecht and Manger explains the method of magnetic attitude control used in the Tiros satellite. An interesting application of the magnetic field for the dumping of angular momentum is described in the paper by McElvain; this scheme has particular application to satellites using an active control system and rotating inertia wheels.

The paper by Williams exhibits a rather complete analysis of the torque and attitude control problem for the syncom satellite, as an example of effects at very great altitudes. The following two papers, by Colombo and Naumann, analyze the observed angular motions of the Explorer XI satellite and deduce its torque experience.

The final set of papers discusses the problem of horizon sensing, which has great importance for satellites requiring earth stabilization and using an active control scheme. A general theoretical discussion is given by Wark, followed by papers by Lunde and Conrath giving particular experience from the Tiros satellite and describing a design for more advanced weather satellites.

A number of the papers in this volume were originally presented at the second Robert H. Goddard Memorial Symposium of the American Astronautical Society, held in Washington, D.C. The editor and authors wish to acknowledge their debt to the AAS, and particularly to the National Capital Section, which organized the Symposium.

S. Fred Singer

October, 1963

Contents

4. Generalized Gravity-Gradient Torques

ROBERT E. ROBERSON

5. Aerodynamic and Radiation Disturbance Torques on Satellites Having Complex Geometry

WILLIAM J. EVANS

6. Forces and Torques Due to Coulomb Interaction with the Magnetosphere

S. FRED SINGER

7. Dynamical Considerations Relating to the West Ford Experiment

R. A. LYTTLETON AND S. FRED SINGER

8. Exploitation of Magnetic Torques on Satellites

RAYMOND H. WILSON, JR.

9. Magnetic Attitude Control of the Tiros Satellites

E. HECHT AND W. P. MANGER

10. Satellite Angular Momentum Removal Utilizing the Earth's Magnetic Field

Robert J. McElvain

11. Torques and Attitude Sensing in Spin-Stabilized Synchronous Satellites

Donald D. Williams

12. On the Motion of Explorer XI around Its Center of Mass

G. Colombo

13. An Investigation of the Observed Torques Acting on Explorer XI

Robert J. Naumann

14. Horizon Sensing in the Infrared: Theoretical Considerations of Spectral Radiance

D. Q. WARK, J. ALISHOUSE, AND G. YAMAMOTO

15. Horizon Sensing for Attitude Determination

BARBARA KEGERREIS LUNDE

16. Earth Scan Analog Signal Relationships in the Tiros Radiation Experiment and their Application to the Problem of Horizon Sensing

BARNEY J. CONRATH

Rotational Motion of Passive Space Vehicles

GORDON S. REITER and W. T. THOMSON

Dynamics Department, Space Technology Laboratories, Redondo Beach, California, and Department of Engineering, University of California, Los Angeles, California

I. INTRODUCTION

IT IS USUALLY DESIRABLE to keep a satellite oriented in a specific attitude. When an inertially fixed attitude is desired, and weight is limited, spin stabilization is often used instead of an active control system.

Over long periods of time, the attitude error of a spin-stabilized satellite will build up because of the cumulative effect of small torques due to magnetic fields and other external effects. Over shorter periods of time, the motion can often be considered force free. The application of force-free motion to current space programs is the subject of this discussion.

The free rotational motion of a spinning satellite relative to its center of gravity depends on the initial conditions. For stability against external torque, a fairly large angular velocity is imparted about a preselected axis during the launch process. Imperfections in the launch process also introduce smaller angular velocity components about the other axes, causing a conical wobbling motion. Wobble can generate irregular fluctuations in the signals from the satellite, or prevent satellites containing cameras from maintaining a steady scan. The design goal is to produce a steady spin about the preselected axis, with no wobble, or at least to keep the initial conditions from becoming worse.

It is possible for a free, spinning satellite, under no external forces, to precess or wobble in a widening cone and finally tumble. The reverse process of damping out a wobble which is initially present is also possible under proper conditions.

The growth or decay of precession from a given initial condition depends on the inertial configuration of the body and on the relative motion of the various parts of the body. A body will be said to be *stable* if a small initial wobble decays and the motion tends toward a steady spin about the intended spin axis. For an *unstable* body, a small initial wobble will grow and the motion will *not* tend toward a spin about the preselected axis.

The conditions governing stability and instability will be derived in the next section. In subsequent sections, these conditions will be applied to the detailed design of wobble-removal devices for stable bodies and to the prediction of wobble buildup in unstable bodies. Examples will be given of the use of such devices on recent space programs.

II. Stability Conditions for Almost-Rigid Satellites; the Energy-Sink Approximation

The motion of a force-free spinning satellite consists of a rotation about the intended spin axis (assumed to be a principal axis) and a roughly conical motion of the intended spin axis about the angular momentum vector, which is fixed in inertial space (Fig. 1). This motion is variously called free precession, nutation, or wobble.

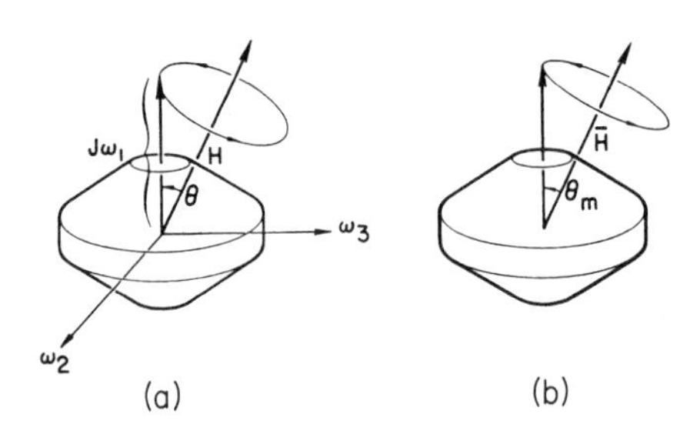

Fig. 1. (a) Free precession of a symmetric satellite. (b) Free precession of an asymmetric satellite.

A steady spin would produce a steady centrifugal stress field in the body. The wobbling motion gives rise to an oscillating stress field superimposed on the steady field.

If the body is not completely rigid, the oscillating stress field set up by the wobble will cause deformation. The resulting conversion of mechanical energy into heat through damping must eventually bring the body into a minimum-energy condition. In this asymptotic condition, the relative motion in the body will vanish. Since the angular momentum vector cannot change for force-free motion, the terminal motion must be a steady spin with some velocity

$$\omega = \frac{H}{I}$$

about some axis having moment of inertia, I, where H is the magnitude of the constant angular momentum vector.

The kinetic energy will then be

$$T_{\min} = \frac{I\omega^2}{2} = \frac{H^2}{2I}$$

The I in the denominator of this expression indicates that for the minimum-energy configuration, I must be the largest moment of inertia which is consistent with the structure of the body. That is, the satellite will eventually change the angle of its coning motion to line up its largest moment of inertia (which may vary in the process for a nonrigid body) along the momentum vector. A *stable* satellite is then one which has its largest moment of inertia about the intended spin axis.

The fact that a nonrigid spinning satellite tends toward a minimum energy state is the basis for all subsequent discussion.

For most applications, it is desirable to predict the way in which the motion tends toward the final stage; that is, the way in which the angle of the cone changes. Such predictions may be made by solving the exact equations of motion, considering the internal degrees of freedom, as discussed in reference [1].

A simpler approximate technique is to assume that the moments of inertia do not vary significantly and that the angular momentum of the relative motion is negligible compared to the rigid-body motion. These assumptions, amply justified in many applications, amount to saying that the motion over any one precession cycle is close to the motion of a completely rigid body having the same energy and angular momentum. The relative motion within the body is then idealized as a slow removal of energy (energy sink) and the rate of energy dissipation can be related to the change in cone angle.

The approximate derivation is especially simple for a symmetric body having moments of inertia J, I, I. The angular momentum magnitude, H, and the kinetic energy, T, may be expressed in terms of the angular velocity components ω_1, ω_2, ω_3 along the body principal axes as

$$H^2 = I^2(\omega_2{}^2 + \omega_3{}^2) + J^2\omega_1{}^2$$

$$T = \tfrac{1}{2}[I(\omega_2{}^2 + \omega_3{}^2) + J\omega_1{}^2]$$

From Fig. 1(a), the relation between the cone half-angle and angular velocities is

$$\sin\theta = I\,\frac{(\omega_2{}^2 + \omega_3{}^2)^{1/2}}{H}$$

Eliminating the angular velocities, we obtain a one-to-one correspondence between cone half-angle and kinetic energy

$$\sin^2\theta = \frac{I}{(J - I)\,H^2}\,(2JT - H^2)$$

or, differentiating with respect to time and assuming constant moments of inertia,

$$2\sin\theta\cos\theta\dot{\theta} = \frac{2IJ}{(J - I)}\,\frac{\dot{T}}{H^2}$$

If θ is to decrease to zero then $\dot{\theta}$ must be negative and, since $\dot{T}$ is negative or zero for an isolated system, θ can only decay if

$$J > I$$

showing again that the terminal condition is spin about the axis of largest moment of inertia. A small conical motion of the intended spin axis can only be damped out if the satellite is stable; that is, if its largest moment of inertia is about the intended spin axis.

The case of

$$\dot{T} = 0$$

corresponds to a completely rigid satellite.

In the case where the body is asymmetric, the intended spin axis moves in an asymmetric cone, rather than a right cone [Fig. 1(b)]. The calculations for this case are more complex, involving elliptic functions [2, 3], but the relation for the maximum half-angle, θ_m, of the cone has a form very similar to that for the symmetric case. It is

$$\sin^2\theta_m = \frac{B}{(A - B)\,H^2}\,(2AT - H^2)$$

where B is the intermediate moment of inertia and A is the moment of inertia about the intended spin axis, either the largest or the smallest. θ_m decreases for negative $\dot{T}$ only if

$$A > B$$

The satellite is stable only if the intended spin axis has the largest moment of inertia.

It may be shown [4] that steady spin about the intermediate axis (having moment of inertia B) is an unstable condition, even for a completely rigid body, and so has no practical application.

The relations above show that the initial wobble of a spinning satellite will be decreased for stable satellites and increased for unstable satellites by removal of kinetic energy through small relative motion within the body, and give the cone angle in terms of the energy dissipation rate. Energy dissipation will depend on the design of the nonrigid parts of the body, which may be fluid or elastic. The design of dampers for intentional energy dissipation in stable satellites will be discussed in the following section, and the design of unstable satellites for minimum dissipation in Section IV.

III. Damping of an Initial Wobble in a Stable Satellite

The various devices used for damping wobble depend on the oscillating acceleration field within the satellite. The acceleration field produces relative motion in the damper, doing work against friction.

The frequency of the oscillating acceleration field at a point within the body (the body-fixed precession frequency) is given [2] by

$$\omega_p = \left(\frac{J}{I} - 1\right) \frac{H \cos \theta}{J}$$

for a symmetric body, and by a similar formula for an asymmetric body. Here I and J are the transverse and spin-axis moments of inertia, respectively, H is the magnitude of the angular momentum vector and θ is the cone half-angle. The forcing frequency is nearly constant for small θ, so that an energy-absorber within the body may be tuned to this frequency to maximize the dissipation.

A typical resonant damping device, analyzed by Taylor in reference [5], is shown in Fig. 2. The device may be analyzed on a computer by means of the exact equations of motion, or approximately by treating the resonant system as a small energy absorber subjected to a prescribed acceleration. Figure 3 shows a computer calculation based on the exact solution for a particular body [5]. The approximate solution leads to an exponential curve similar to the one shown, but without the oscillations.

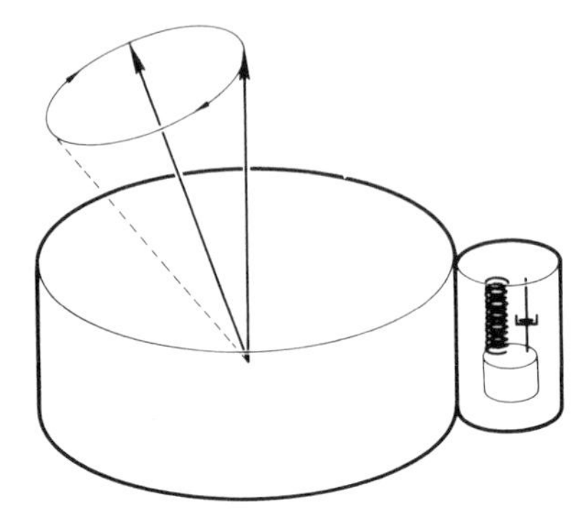

FIG. 2. Spinning satellite with simple damping device.

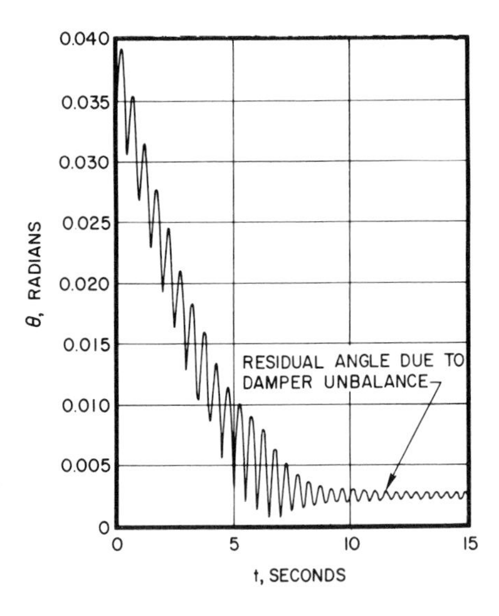

FIG. 3. Cone angle decay due to resonant energy absorber.

Several resonant systems, identical in principle to the one described above, have been studied [6-8]. One such system is the Team damper used by RCA on the Tiros program [6]. The Team configuration is essentially a centrifugal pendulum, in which centrifugal force replaces the mechanical spring in the single-degree-of-freedom system.

Resonant mechanical systems are very effective at moderate values of the cone angle, but usually have some static friction which causes them to stop moving at low stress levels. Devices which dissipate energy through viscosity in a liquid have been used for applications in which the cone angle is small.

The usual liquid-damper configuration, originated at the Naval Ordnance Test Station, is shown in Fig. 4(a). The device consists of an

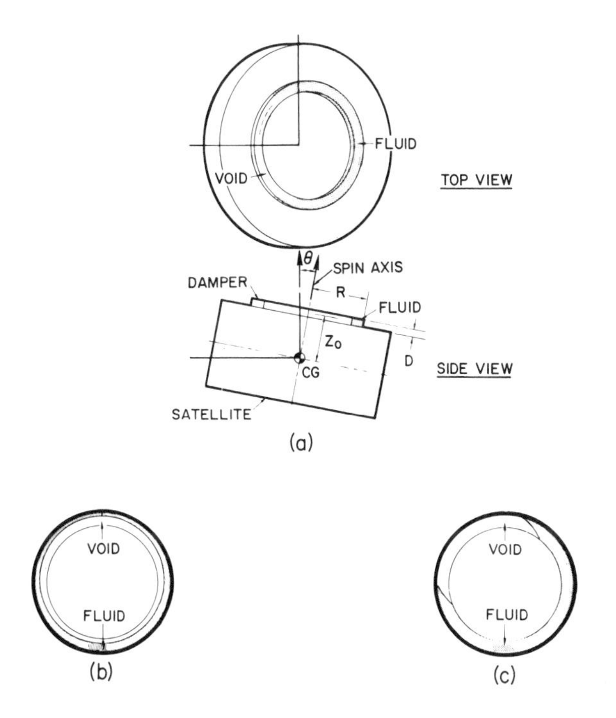

FIG. 4. (a) Annular damper mounted on a spin-stabilized body. (b) Annular damper configuration at low cone angles. (c) Annular damper configuration at high cone angles.

annulus partially filled with liquid. The annulus is mounted concentric with the spin axis, at a point off the axial center of gravity.

At sufficiently high cone angles, the unbalanced centrifugal acceleration due to precession keeps the liquid in a solid slug on one side of the annulus (Fig. 4(c)) and the body spins past it. The motion is very similar to the flow of viscous liquid in a pipe, and has been analyzed on that basis [9, 10].

Figure 5 [9] shows a comparison between turbulent pipe flow theory and experiment for a laboratory model using mercury as the liquid. The ratio of spin axis to transverse axis moment of inertia was 1.56 and the mercury mass was roughly 1/1000 of the body mass. The annulus was one-fourth full of mercury. The experiments were conducted at Space Technology Laboratories (STL) by Fitzgibbon and Smith.

Agreement with theory was good in general, as long as the mercury remained in a slug. As the angle became smaller, the liquid gradually spread out in an annulus with a free surface (Fig. 4(b)), and the theory ceased to apply, as Fig. 5 shows.

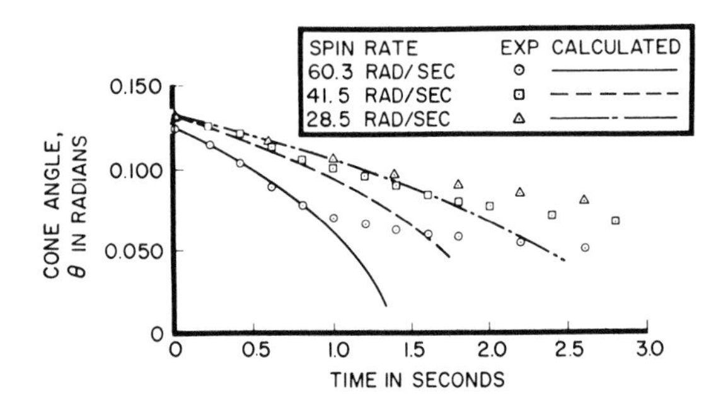

Fig. 5. Cone angle decay with annular damper—comparison of theory and experiment.

The range of cone angles at which the transition from pipe flow to free surface flow will occur for a given body is determined by the relationship between viscous forces and inertial forces.

In the spread-out configuration, the annular damper resembles an open channel with surface waves. This model has been used to analyze the device [11].

The analysis predicts a resonance, or maximum-dissipation condition, when the lowest surface-wave frequency coincides with the body-fixed precession frequency.

It is logical to exploit the resonance condition to obtain fast damping. However, the STL experiments show that, even for damper weights on the order of 2 % of the main body weight, the annular damper can store a significant amount of energy in surface waves, and then feed it back to the body, rather than removing it. This is illustrated in Fig. 6 [9], which shows measurements of the cone angle for a laboratory model near the predicted damper resonance. The transfer of energy back and forth between rigid-body motion and surface-wave motion is evident.

For the parameters used in the tests, the feedback mechanism essentially had the effect of removing the dependence of damping on frequency, so that no increase in damping was obtained by tuning the damper frequency to the body precession frequency. Some additional dissipation was obtained by filling the annulus sufficiently full so that the surface waves in the fluid could impact on the inside surface of the annulus, and also by filling the void in the annulus with a light liquid.

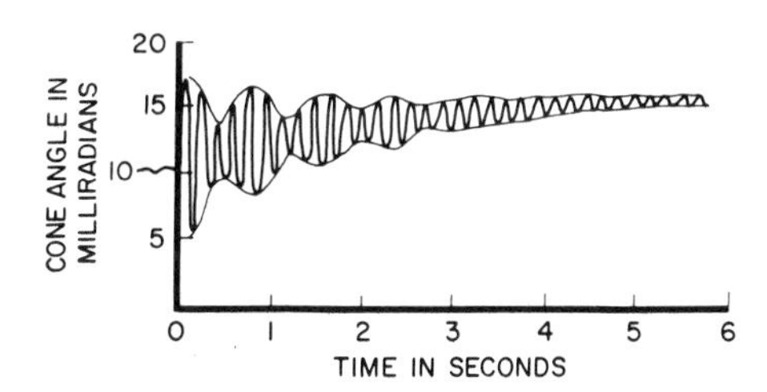

FIG. 6. Measured cone angle change near resonance with annular damper.

Because of the energy-transfer phenomenon and other effects, there is presently no adequate way to predict the performance of annular dampers without testing, except for pipe-flow theory, which applies if the cone angle is large enough.

Properly used, annular dampers can be very effective. In some of the STL tests, cone angles were reduced by factors of 5 or more in five spin cycles.

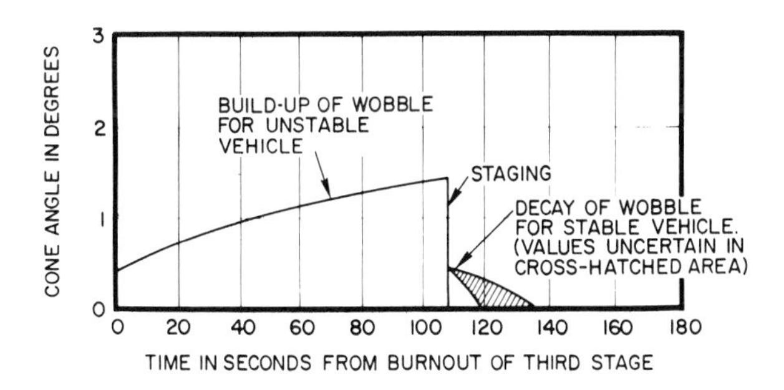

FIG. 7. Explorer VI launch vehicle: measured cone angle change due to annular damper.

In these tests, the moment-of-inertia ratio was 1.55 and the mercury damper weight was roughly 1/60 of the body weight.

As an example of the use of annular dampers on satellite vehicles, Fig. 7 shows an accelerometer measurement of the cone half-angle as a function of time for the Explorer VI launch vehicle. During the first phase, the configuration is unstable and an annular damper is building up the wobble. After a change to a stable moment-of-inertia ratio at staging, the wobble rapidly decays.

IV. Control of Wobble Buildup in unstable Satellites

It is often necessary to allow a spinning vehicle to remain in an unstable moment-of-inertia configuration for some appreciable period of time. For example, in the Pioneer V launch, the vehicle had to coast for 20 minutes after third-stage burning, in an unstable configuration, until the third-stage rocket case could be separated.

When an unstable configuration is required, it is necessary to design the vehicle to minimize internal motion.

To calculate the wobble increase that will occur in an unstable satellite, the structure is idealized as being all or partly elastic. Hysteresis damping due to solid friction is included. The elastic deformation is computed from the inertia forces due to the coning motion. In each precession cycle, a small fraction of the energy stored in elastic deformation is converted to heat, and produces a corresponding change in the cone angle. The resulting divergence of the cone angle is exponential in time.

The calculations decribed above have been carried out for several satellite configurations, such as a disk [12], a solid body with projecting beams [13], and two rods connected by a torsional spring [14]. Unfortunately, structural damping is not the only mechanism for energy dissipation. Damping in structural joints, in bending wires, and in battery fluid may be important. Analyses based on an idealized structure give only a rough lower bound to the wobble increase, and must be treated with caution.

Very little flight data on unstable satellites is available to the authors. The pencil-shaped Explorer I [15] had flexible turnstile antennas protruding from its sides. The antennas acted as centrifugal pendulums resonant at a frequency very near the precession frequency. The resulting dissipation caused appreciable tumbling in less than one orbital revolution. Explorer III, having a similar configuration without the antennas, took several days to build up a large wobble angle.

The Pioneer V payload and third stage, typical of the Able and Thor-Delta class of vehicles, coasted for 20 minutes (about 2000 precession cycles) in an unstable configuration with no detectable change in the initial cone angle of 1 or 2 deg.

Acknowledgment

The authors gratefully acknowledge the assistance of R. S. Taylor and of D. P. Fitzgibbon and W. E. Smith, who made available the experimental results described above.

References

1. R. E. Roberson, Torques on a satellite vehicle from internal moving parts. *J. Appl. Mech.* **25**, 196, 287 (1958).
2. W. T. Thomson, "Introduction to Space Dynamics," p. 126. Wiley, New York 1961.
3. M. E. Kuebler, Gyroscopic Motion of an Unsymmetrical Satellite Under no External Forces, *NASA Tech. Note D-596*, December 1960.
4. R. F. Deimel, "Mechanics of the Gyroscope," p. 68. Dover, New York, 1950.
5. R. S. Taylor, A Spring-Mass Damper for a Spin-Stabilized Satellite, *Space Technol. Lab. Rept. EM* **11-15**, July 1961.
6. H. Perkel, Tiros I spin stabilization. *Astronautics* **5**, 38-39, June 1960.
7. S. J. Zoroodny and J. W. Bradley, Nutation Damper—A Simple Two-Body Gyroscopic System, *Ballistic Research Lab. Rep.* **1128**, April 1961.
8. H. L. Newkirk, W. R. Haseltine, and A. V. Pratt, Stability of rotating space vehicles. *Proc. I.R.E.* **48**, 743-750 (1960).
9. D. P. Fitzgibbon and W. E. Smith, Final Report on Study of Viscous Liquid Passive Wobble Dampers for Spinning Satellites, *Space Technol. Lab. Rep. EM* **11-14**, June 1961.
10. E. E. Rogers, A Mathematical Model for Predicting the Damping Time of a Mercury Damper, *U.S. Naval Ordnance Test Sta. Rept. No. IDP* **565**, 2 March 1959.
11. G. F. Carrier and J. W. Miles, On the annular damper for a freely precessing gyroscope. *J. Appl. Mech.* **27**, 237-240 (1960).
12. L. Meirovitch, Attitude Stability of a Disk Subjected to Gyroscopic Forces. Ph. D. Thesis, Univ. of California, Los Angeles, 1960.
13. W. T. Thomson and G. S. Reiter, Attitude drift of space vehicles. *J. Astronaut. Sci.* **7**, 29-34 (1960).
14. H. Perkel, Effect of Energy Loss during Coast Period of Combined Third Stage Rocket and Tiros Payload, *Radio Corp. of America, Tiros TM 232-19*, 29 June 1959.
15. W. C. Pilkington, Vehicle Motions as Inferred from Radio-Signal-Strength Records, *Jet Propulsion Lab. Extern. Publ. No. 551*, September 5, 1958.

Passive Gravity-Gradient Stabilization for Earth Satellites

ROBERT E. FISCHELL

The Johns Hopkins University, Applied Physics Laboratory, Silver Spring, Maryland

I. INTRODUCTION

GRAVITY-GRADIENT ATTITUDE STABILIZATION is defined as the alignment of one axis of a satellite along the earth's local vertical direction so that a particular end of the satellite always faces in the downward direction. Passive gravity-gradient stabilization is defined as achieving this orientation, including damping the resulting librations, without the use of active control elements such as servo systems, reaction wheels, or gas jets. Passive techniques can include moving parts which utilize the environment of the satellite (such as the gravity-gradient itself) to damp oscillations about the local vertical.

There are numerous advantages for satellites that are aligned vertically with the same side continually facing downward. Probably the greatest advantage is that a directional satellite antenna can be utilized to enhance the signal strength of radio transmission both to and from the orbiting satellite. In a similar manner, improved optical tracking of a brilliant flashing light, such as used on the Anna satellite, can be obtained by directing the light only in a downward direction. The power gain (compared to an isotropic radiator) that can be realized by means of a gravity-stabilized satellite which sends all its radiation to cover the earth from horizon to horizon is shown in Fig. 1. At the altitude of a synchronous satellite (24-hour orbital period) gravity stabilization is almost a requirement for effective communication to and from the satellite.

Gravity-gradient stabilization offers many advantages for earth observations by means of cameras on an orbiting spacecraft. For example, the quantity of usable pictures produced from a meteorological or surveillance satellite can be considerably increased when the cameras are always directed toward the earth's surface.

Several scientific experiments to study corpuscular and electromagnetic radiation are more profitably performed on a vertically oriented spacecraft.

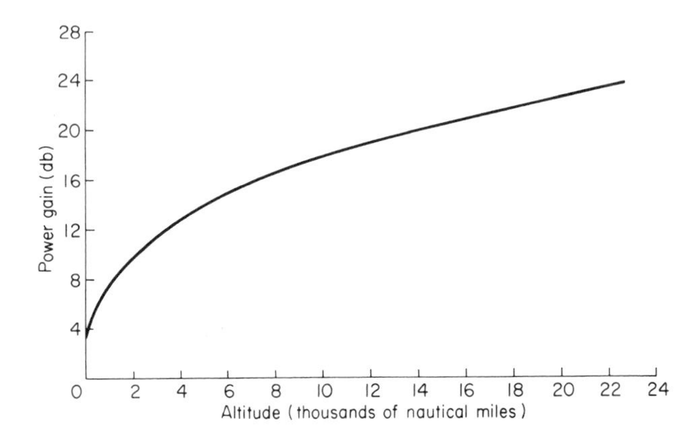

FIG. 1. Power gain at the surface of the earth for a gravity-oriented satellite with a directional transmitting pattern covering from horizon to horizon.

A less obvious advantage is that gravity stabilization can be employed to improve the operation of solar cell power generating systems. With a system of gravity stabilization, it is possible to design the satellite so that the projected area of the solar cells is inversely proportional to the percent solar illumination [1]. This assures a constant rate of electrical power generation irrespective of the fraction of the time that the satellite is in the sun. This same principle of a change in projected area depending on the percentage of time the satellite is illuminated by the sun can be applied to enhance the thermal design of a satellite [1].

II. Theory of Gravity-Gradient Attitude Stabilization

The principle on which gravity-gradient attitude stabilization is based is quite simple. A considerable number of theoretical papers have been written on the subject [2-4]. Putting these ideas into practice is however, a most difficult engineering problem. In Figure 2 is shown a satellite which is essentially in the shape of a "dumbbell." For the purpose of this discussion let us assume that the center of gravity of the system is contained within the major satellite instrumentation in the section marked A in Fig. 2. Extended a considerable length outward from the satellite's center of gravity is a mass, B in Fig. 2. The center of gravity of a satellite is in a stable orbit when the centrifugal force due to the satellite motion in a curved trajectory is precisely equal, but in the opposite direction, to the gravitational force resulting from the attraction of the mass of the earth.

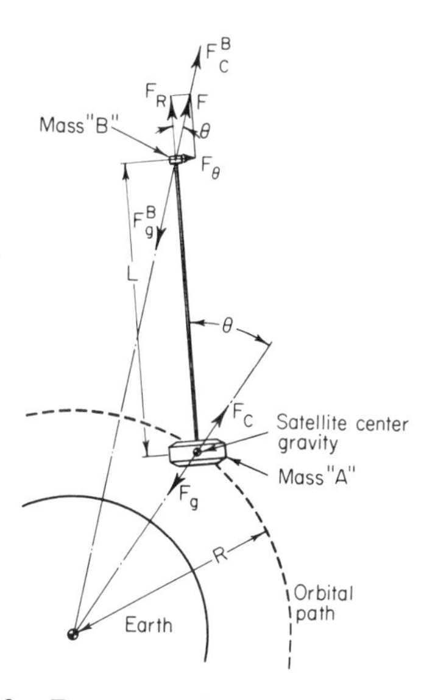

FIG. 2. Forces on a dumbell-shaped satellite.

If we let F_g be the gravitational force at the center of gravity of the satellite, and F_c be the centrifugal force at that point, then a stable orbit is defined by

$$F_g = F_c \tag{1}$$

and for $(L/R)^2 \ll 1$ (on the order of 10^{-10} for a reasonably dimensioned satellite)

$$F_g = \frac{GM(m_A + m_B)}{R^2} \qquad \text{(dynes)} \tag{2}$$

and

$$F_c = (m_A + m_B)\,\omega_0^2\,R \qquad \text{(dynes)} \tag{3}$$

where

G = gravitational constant ($cm^3\ gm^{-1}\ sec^{-2}$)
M = mass of the earth (grams)
m_A = mass of the main section of the satellite (grams)
R = orbital radius (cm)

and

ω_0 = orbital angular rate (radians/sec)

When

$$\frac{d\theta}{dt} = 0$$

the forces on the mass B are given by

$$F_g{}^B = \frac{GMm_B}{(R + L\cos\theta)^2} \qquad \text{(dynes)} \tag{4}$$

$$F_c{}^B = m_B\omega_0{}^2(R + L\cos\theta) \qquad \text{(dynes)} \tag{5}$$

where L = distance from the satellite center of gravity to the mass B (cm), and θ = angle of the satellite symmetry axis with respect to the local vertical (degrees).

Since $F_g = F_c$ defines a stable orbit, then

$$F_c{}^B > F_g{}^B \tag{6}$$

and there is a net outward force on mass B given by

$$F = m_B\left[\omega_0{}^2(R + L\cos\theta) - \frac{GM}{(R + L\cos\theta)^2}\right] \qquad \text{(dynes)} \tag{7}$$

The component along the length of the boom causes a tension in the structural member attaching the mass B to the satellite body. When $d\theta/dt = 0$, this tension force is given by

$$F_R = m_B\left[\omega_0{}^2(R + L\cos\theta) - \frac{GM}{(R + L\cos\theta)^2}\right]\cos\theta \qquad \text{(dynes)} \tag{8}$$

When $d\theta/dt \neq 0$, there is an additional contribution to F_R due to the angular rate of the mass B about the center of gravity of the satellite.

The force perpendicular to the symmetry axis provides the restoring torque given by

$$\tau = F_\theta L = m_B\left[\omega_0{}^2(R + L\cos\theta) - \frac{GM}{(R + L\cos\theta)^2}\right]L\sin\theta \qquad \text{(dyne-cm)} \tag{9}$$

It can also be shown [4] that for a dumbbell-shaped satellite with moments of inertia I_x and I_z, the torque is given by

$$\tau = \frac{3}{2}\,\omega_0^2(I_x - I_z)\sin 2\theta \qquad \text{(dyne-cm)} \tag{10}$$

For simplicity it is assumed that the satellite has cylindrical mass symmetry; i.e., $I_x = I_y$. To develop a substantial torque it is necessary that I_x be very much greater than I_z. We also see that the gravity-gradient torque is less effective for satellites at very high altitudes where the orbital period is very great and therefore ω_0^2 is very small.

The natural period of oscillation (libration period) of a gravity-stabilized satellite is given by

$$T_{\parallel} = \frac{2\pi}{\omega_0\sqrt{3(1 - I_z/I_x)}} \qquad \text{(seconds)} \tag{11}$$

in the plane of the orbit; and by

$$T_{\perp} = \frac{\pi}{\omega_0\sqrt{1 - I_z/I_x}} \qquad \text{(seconds)} \tag{12}$$

in the plane perpendicular to the orbit. For a satellite having an orbital period of 100 minutes and having $I_x \gg I_z$ we find that $T_{\parallel} = 57.8$ minutes and $T_{\perp} = 50.0$ minutes. These very long libration periods, when combined with the trivial torques that are available, make damping of the satellite oscillations a most difficult problem.

III. Procedure for Achieving Stabilization

For an earth satellite to achieve passive gravity-gradient attitude control, it is necessary to follow certain procedures. These procedures will of course differ somewhat for various satellite missions, but some problems common to all will be discussed herein. It should be presumed that the long extension, or boom, that is required to alter the mass distribution of the satellite will be extended after the satellite is in orbit.

The first thing that must be accomplished is to remove virtually all the spin that may have been imparted to the satellite during the launch procedure. A device which rapidly removes the spin energy of a satellite is the so-called "yo-yo" consisting of two weights attached to cables which are wrapped around the satellite [5]. When the weights are released they spin out from the satellite causing a tension in the cables which results in a retarding torque on the satellite. This device has been successfully employed on navigational satellites as well as on several Tiros satellites. To guarantee the very low angular rates that are required for erecting a comparatively weak extendible boom, one can employ magnetic hysteresis rods [6]. By rotating in the earth's magnetic field these rods remove the spin energy of the satellite because of their magnetic hysteresis loss. Magnetic damping has been successfully employed for removing the spin energy of the Anna and navigational satellites. An early navigational satellite employed both "yo-yo" and magnetic despin devices to eliminate unwanted spin [5]. For the first 7 days, magnetic rods mounted perpendicular to the satellite's spin axis created a retarding torque by magnetic hysteresis loss reducing the spin rate from 2.80 to 2.60 rps. On the seventh day the "yo-yo" despin weights were deployed reducing the spin rate from 2.6 rps to approximately 0.08 rps. The magnetic rods reduced the remaining spin to approximately two revolutions per orbit (rpo) in a period of less than 10 days. This final spin rate of less than 0.001 rps is quite slow enough to allow the deployment of a comparatively weak extendible boom.

The next procedure is to align the satellite vertically with the correct side facing downward. This can be accomplished by energizing an electromagnet internal or rigidly attached to the satellite. The direction of the resulting magnetic dipole moment is along the satellite's Z (symmetry) axis. It can then be shown that the satellite will align its Z axis along the local magnetic field direction [7]. The magnetic hysteresis rods that were used to remove the spin energy of the satellite will also damp the oscillations of the satellite about the local magnetic field direction [7]. A magnetically stabilized satellite over the earth's magnetic pole will be stabilized along the local vertical (which is the direction of the local magnetic field) with a *particular*, predetermined face of the satellite directed downward. It can be shown that the tumbling rate of the satellite at this time will be 1.5 rpo [7]. The satellite will now be in a most advantageous condition for capture into gravity-gradient attitude stabilization.

The boom will then be erected and the electromagnet turned off by radio command from a ground station. The satellite will then have its tumbling angular rate reduced by the ratio of the satellite's moment of inertia after the erection of the boom compared to the moment of inertia before. For a typical satellite design, the moment of inertia might be increased by a factor of 100, resulting in a decrease in the satellite's tumbling rate to 0.015 rpo, which is essentially stopped in inertial space. In order to be vertically stabilized the satellite must then achieve a tumbling rate in inertial space of 1.0 rps. Immediately after the boom is erected the satellite will continue in its orbital motion with its Z axis essentially fixed in inertial space. As the satellite moves away from the magnetic pole, a gravity-gradient force will act upon it tending to align the Z axis along the local vertical direction. The angle with the local vertical will continue to increase until the gravity-gradient torque causes the satellite to develop an angular rate of 1.0 rpo. The satellite angle with the vertical will then decrease as the gravity-gradient torque continues to act, resulting in a planar libration motion of the satellite. This is illustrated in Fig. 3.

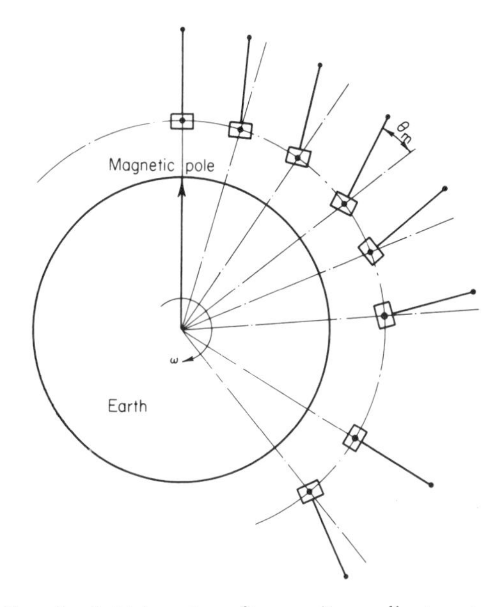

FIG. 3. Initial motion after gravity-gradient capture.

The maximum angle developed with the local vertical is of great interest. If this angle is less than 90 deg, capture of the satellite into gravity-gradient attitude stabilization will result. The angle can be calculated rather simply by equating the angular kinetic energy that the satellite must develop (to achieve an angular rate of 1.0 rpo) to the work done by the gravity-gradient torque as the satellite moves out to that maximum angle.

The satellite angular kinetic energy when it achieves an angular rate 1.0 rpo is given by

$$\text{K.E.} = \tfrac{1}{2} I_x \omega_0^2 \qquad \text{(ergs)} \tag{13}$$

The work done on the satellite by the gravity-gradient torque is given by

$$W = \int_{\theta_i}^{\theta_m} \tau d\theta \qquad \text{(ergs)} \tag{14}$$

where θ_i = initial angle off the vertical, and θ_m = the maximum angle with the local vertical direction to which the satellite will swing.

Taking the expression for τ from Eq. (10), and setting $\theta_i = 0$, gives

$$W = \tfrac{3}{2}\,\omega_0^2(I_x - I_z) \int_0^{\theta_m} \sin 2\theta \, d\theta \tag{15}$$

$$W = -\tfrac{3}{4}\,\omega_0^2(I_x - I_z)\,(\cos 2\theta_m - 1) \tag{16}$$

Equating (13) and (16) and solving for θ_m yields the result

$$\theta_m = \tfrac{1}{2} \operatorname{arc\,cos} \left[1 - \tfrac{2}{3}\left(\frac{I_x}{I_x - I_z}\right)\right] \tag{17}$$

A practical design for gravity stabilization requires that $I_x \gg I_z$; therefore, we can set

$$\frac{I_x}{I_x - I_z} \cong 1 \tag{18}$$

which gives

$$\theta_m = \tfrac{1}{2} \operatorname{arc\,cos} 0.333 \tag{19}$$

or

$$\theta_m = 35.36 \text{ degrees}$$

Since θ_m is well below 90 deg, the satellite will be captured into the vertical stabilization condition. Furthermore, the maximum angle is independent of the orbital period (and therefore independent of the satellite altitude). For $I_x \gg I_z$, the maximum angle is also independent of the value of I_x. If the satellite is not exactly aligned along the local field direction when the boom is erected, the satellite will swing out to a larger angle with respect to the local vertical direction. For $\theta_i \neq 0$ it can readily be shown that the peak angle to which the satellite will swing before achieving an angular rate of 1.0 rpo will be given by

$$\theta_m = \tfrac{1}{2} \operatorname{arc\,cos} \left[\cos 2\theta_i - \tfrac{2}{3}\left(\frac{I_x}{I_x - I_z}\right)\right] \tag{20}$$

The result of θ_m as a function of θ_i for $I_x \gg I_z$ is shown in Fig. 4. From Fig. 4 it can be seen that θ_m increases as the absolute value of θ_i increases;

i.e., for positive or negative initial deviation angles off the vertical, the maximum angle θ_m will increase. The limiting angle for capture, $\theta_m = 90$ deg, occurs at $\theta_i = 54$ deg.

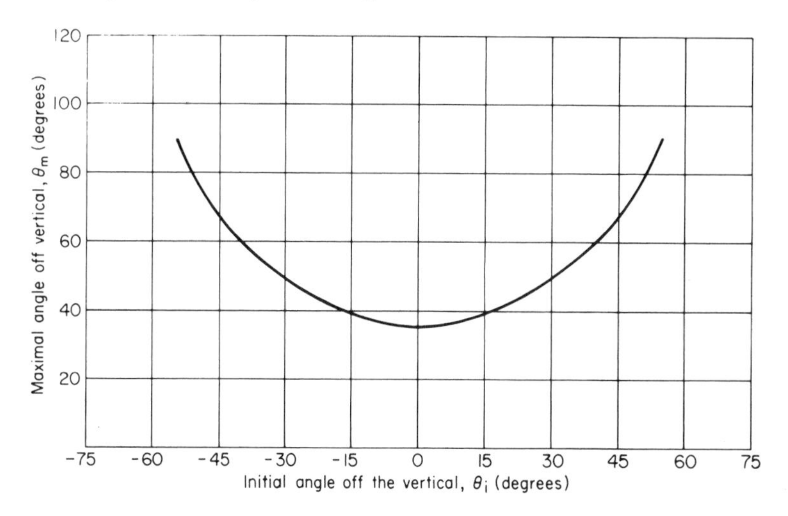

FIG. 4. Maximum angle off the vertical as a function of initial angle.

The previous discussion describes the capture process for a satellite whose orbit takes it over one of the earth's magnetic poles. This would include orbits with inclinations between 74 and 90 deg for capture over the north magnetic pole and between 68 and 90 deg for capture over the south magnetic pole. This capture process can be usefully performed using the above-described process for satellites with a considerably lower inclination, but oscillations perpendicular to the orbital plane will result. These oscillations will then have to be damped out just as the initial 35.36 deg oscillation in the plane of the orbit should be damped out. The technique of initial alignment along the earth's magnetic field is quite practical for orbital inclinations as low as 30 deg. At 30 deg N latitude, (and 75 deg E longitude) a magnetically stabilized satellite would only be 27 deg off the local vertical direction. After the boom is extended, a satellite capture at this point would have an initial cross-orbit oscillation of 27 deg as well as the initial oscillation in the plane of the orbit of 35.36 deg.

An alternate scheme to that described above would be to allow the satellite's Z axis to drift in a random manner and merely wait to erect the boom by a command from some station when the Z axis was observed to be within some acceptable angle (*viz.* 30 deg) with respect to the local vertical.

The next procedure, damping of the satellite librations, is undoubtedly the most difficult that must be accomplished to achieve gravity-gradient stabilization. Damping is required to remove initial librations and to reduce the effect of perturbing torques and impulses. Several methods have been suggested for this purpose [8, 9]. There are undoubtedly numerous possibilities for achieving damping. A most promising method, the use of an energy-dissipating spring, was suggested by Dr. R. R. Newton of the Applied Physics Laboratory. For the Traac

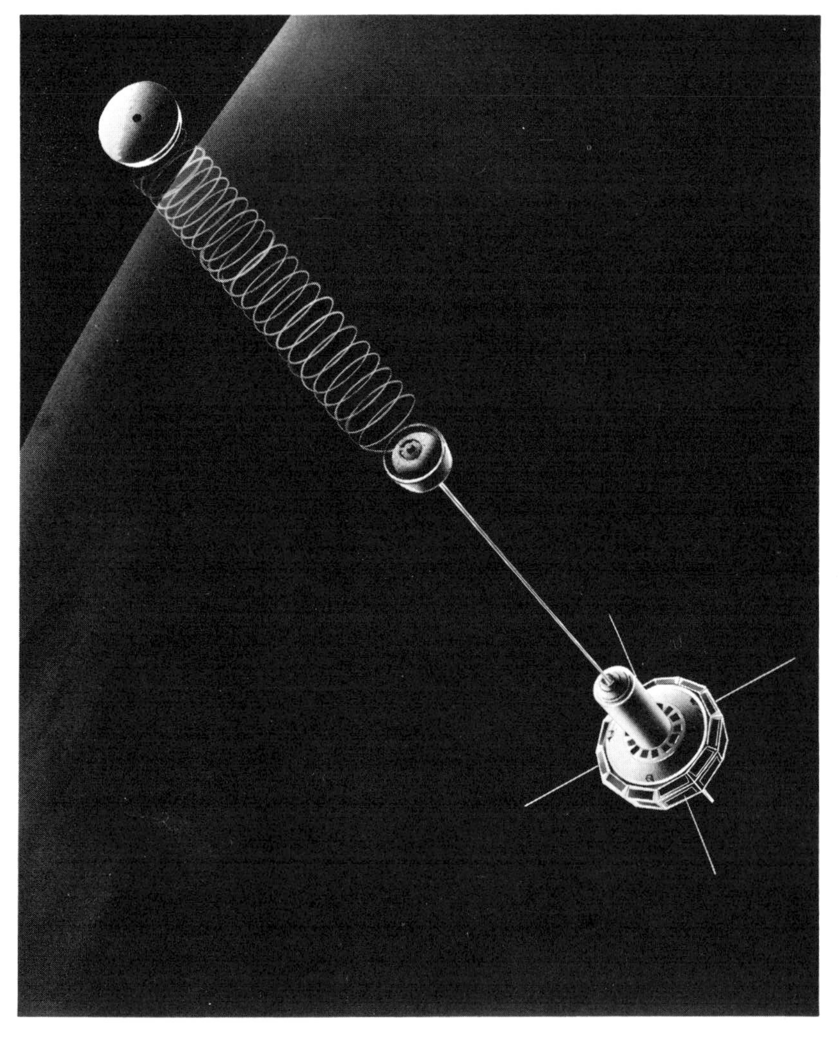

FIG. 5. The Traac satellite.

satellite, damping of the libration motion was to be accomplished by this ultraweak spring fastened to the end of a long boom [10].

Figure 5 is an artist's conception of the orbiting Traac satellite with the boom and spring extended. The spring, with its associated end mass, is held rigidly to the end of the boom during the launch and magnetic stabilization phases by means of a block of subliming material. After the boom is erected this material initiates its sublimation. First to be released is the mass that it attached to the end of the spring. The gravity-gradient force then acts on this mass and tends to pull the spring out. The spring is packed in subliming material so that one coil is released at a time. Although the spring has zero length in its equilibrium position in a zero g field, the gravity-gradient force acting on the end mass was to cause it to have a nominal length of approximately 40 feet owing to gravity-gradient force.

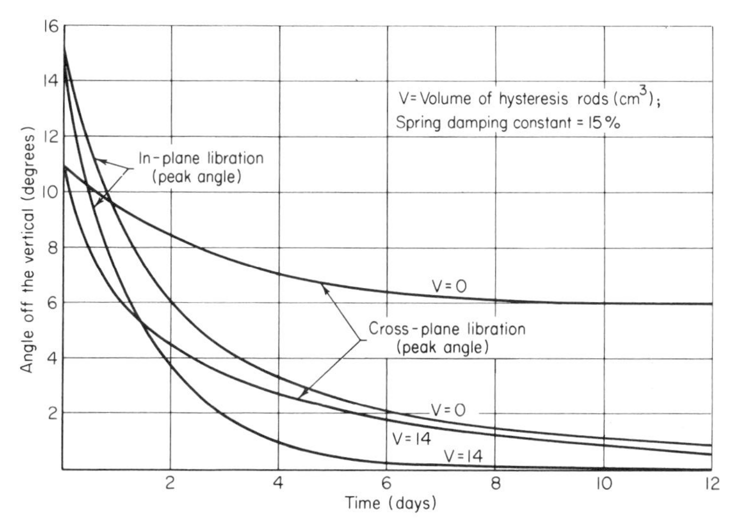

FIG. 6. Damping of satellite librations for a satellite with damping spring and hysteresis.

As the satellite oscillates about the local vertical, the radial force in the direction of the boom varies owing to the difference in gravity-gradient force as a function of the angle θ plus an additional force contribution due to the $d\theta/dt$ term originating from the librational motion. This varying force causes the spring to move in and out. The libration energy is then absorbed by mechanical hysteresis in the spring.

Dr. J. L. Vanderslice of the Applied Physics Laboratory, and Dr. B. Paul of the Bell Telephone Laboratories [11] have analyzed the motion of a boom-and-spring system. The results of both investigators determined that the spring provides the damping required for gravity-gradient stabilization. The analysis of Dr. Vanderslice has shown that the spring is extremely effective in damping satellite librations in the plane of the orbit and is less effective in damping oscillations perpendicular to the orbital plane. The magnetic rods have some damping effect for motions in the orbital plane and are most effective for damping the cross-plane oscillations. Figure 6 shows the effectiveness in damping satellite librations for a satellite with and without magnetic damping rods.[1] The effectiveness of the spring and rods for damping all satellite librations is quite apparent from these curves.

IV. Design of the Damping Spring

The fabrication of a damping spring and the design of a simple but reliable means of deploying it in orbit present a difficult engineering problem. Typical of the springs that might be employed for this purpose was the one used on the Traac satellite. When extended, this damping spring was to have the shape of the surface of a frustum of a cone. In its equilibrium position the spring was annealed to lie flat. In this position the inside diameter of the spring is 4 inches and the outside diameter of the spring is $7\frac{1}{4}$ inches. The spring consisted of 142 turns of 0.007-inch diameter beryllium copper wire. Since beryllium copper is an "excellent" spring material, it does not provide sufficient energy absorption. To obtain good damping, a 0.0008-inch thick layer of the mechanically soft material cadmium was electrolytically deposited on the outer surface of the beryllium copper wire. Since cadmium tends to sublimate in the hard vacuum of space a 0.0002-inch coating of silver was electrolytically deposited on the outer surface of the cadmium to prevent this possibility. When completely fabricated, annealed and coated, the spring had a constant of $1\frac{1}{2} \times 10^{-6}$ pounds per foot.

To determine the effectiveness of the spring under the conditions expected in orbit, a torsional pendulum using the spring material was built and tested in a large vacuum chamber. The period of the torsional pendulum was set at 55 minutes to closely correspond to the natural period of libration for the nominal orbit of the Traac satellite. By this method it was determined that the spring used on the Traac satellite had a damping coefficient of 15 %; i.e., 15 % of the maximum energy stored in the spring was dissipated on each oscillation. This compares

[1] These curves were obtained by Dr. Vanderslice employing an IBM 7090 computer to analyze the motion of a spring-damped, librating satellite.

with an energy loss of less than 1 % per cycle for the uncoated beryllium copper wire. More recent coated springs have achieved an energy loss per cycle in excess of 50 %.

To prevent any tangling or other damage to this ultraweak spring during handling and launching operations it was necessary to encapsulate the spring in a solid subliming material. The material selected for this purpose was a compound of the benzine family known as biphenyl. After the Traac satellite spring was formed it was placed on a conically shaped holder and the molten biphenyl was poured on top of it to assure complete encapsulation. The biphenyl was also used to securely hold the spring end mass during the erection of the boom. After the boom was extended the biphenyl was to sublime away; first releasing the end mass, and further sublimation of the biphenyl was to allow one coil of the spring to extend at a time. Complete deployment of the spring from the subliming material was to be accomplished within a period of 2 days.

Although the Traac satellite boom did not deploy, the spring and its associated end mass apparently did so to a limited extent. To verify this, a deliberate rocking motion about the local magnetic field direction was induced by means of the electromagnet in the satellite. Prior to the deployment of the spring, the satellite magnetic damping rods would damp these oscillations about the local magnetic field direction from approximately 60 deg to 10 deg in a 5-day period. However, after the spring deployed, the satellite oscillations damped from 60 deg to less than 10 deg in one orbital period of approximately 100 minutes. This high rate of damping could only have been caused by the damping action of the spring. This verified in orbit the deployment technique and high energy absorbing characteristics of the spring.

The spring was last checked in June 1962, 7 months after launching, and was found to have retained its high damping characteristics. This test provided assurance that the cadmium coating had not sublimed from the spring.

V. Effect of Perturbing Torques

It has been shown that a satellite having $I_x \gg I_z$ and with an effective damping spring and hysteresis rods will stabilize vertically along the direction of the earth's gravity gradient. To assure that this is the case for a practical satellite design one must examine all possible torques that tend to perturb the satellite off its vertical position. The principle perturbing torques are (1) magnetic, (2) solar radiation pressure, and (3) aerodynamic. Magnetic interactions are a principal perturbing torque for satellite altitudes below 1000 miles altitude; solar radiation pressure would be the most significant disturbing effect for a satellite in a synchronous orbit; the effect of aerodynamic drag is relatively unimportant for satellite altitudes above 300 miles.

The magnetic torque is a result of the interaction of the earth's magnetic field with any permanent or induced dipole moment from permeable material in the satellite. When hysteresis rods are employed for damping they are usually the largest source of magnetic dipole moment. A typical satellite design employing magnetic damping rods would produce a residual magnetic dipole moment on the order of 10^3 unit-pole cm. The torque resulting from the interaction with the earth's magnetic field is given by

$$\tau_m = MH \sin \varphi \qquad \text{(dyne-cm)} \tag{21}$$

where

M = satellite's magnetic dipole moment (unit-pole cm)

H = earth's magnetic intensity at the satellite (oersted)

φ = angle between the earth's magnetic field and the magnetic dipole of the satellite.

A typical value for H at 500 miles altitude is 0.3 oe. This gives a maximum torque (based on a dipole moment of 10^3) of 300 dyne-cm. It is readily possible to achieve moments of inertia for the satellite (*viz.*, the Traac satellite) such that $I_x - I_z \cong 10^{10}$ gm-cm^2. From Eq. (10) we find, for a satellite altitude of 500 miles ($\omega_0 \simeq 10^{-3}$), that the gravity-gradient torque would be given by

$$\tau = 1.5 \times 10^4 \sin 2\theta \qquad \text{(dyne-cm)} \tag{22}$$

If the magnetic torque were applied continuously this would result in a deviation off the vertical given by

$$\theta = \tfrac{1}{2} \arcsin \frac{300}{1.5 \times 10^4} \tag{23}$$

or

$$\theta = 0.57 \text{ degrees}$$

Since the earth's magnetic field intensity varies as the inverse cube of the distance from the center of the earth, orbits considerably higher than 500 miles altitude will produce a significantly smaller magnetic perturbing torque.

For a gravity-stabilized satellite with an asymmetric distribution of area about the center of gravity there will be a net perturbing torque due to solar radiation pressure. Figure 7 illustrates the configuration of the Traac satellite. For this satellite one can consider the center of gravity to be located within the instrumentation section indicated by mass A in

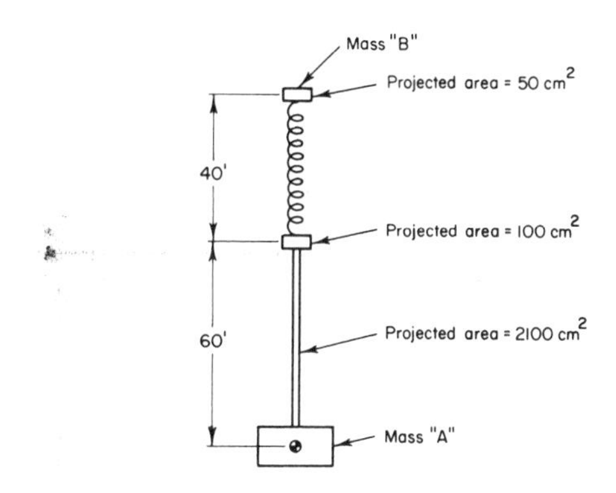

FIG. 7. Configuration of the Traac satellite.

Fig. 7. The torque resulting from solar radiation pressure is given by

$$\tau_s = (1 + C_r)\, A_p P_s d \qquad \text{(dyne-cm)} \tag{24}$$

where

C_r = that fraction of incident photons reflected from the surface (one assumes for this case that all others will be absorbed)

A_p = the projected area (cm^2)

and

P_s = the radiation pressure exerted by the sun on a totally absorbing surface (this will be taken here as 4.5×10^{-5} dynes/cm^2)

d = the distance from the center of gravity of the satellite to the center of pressure for that surface (cm)

For the Traac satellite as illustrated in Fig. 7, one can calculate the perturbing effect of solar radiation pressure assuming the boom to have $C_r = 0.5$, and the surfaces at the end of the boom and the spring to have a $C_r = 0$. The results of these calculations indicate that the solar torque would be approximately 150 dyne-cm. Comparing this to the 300 dyne-cm perturbing magnetic torque indicates that the solar torque will cause approximately $\frac{1}{4}$-deg displacement of the satellite off the local vertical direction.

Although this effect was small for the Traac configuration, the use of much larger booms to gravity-stabilize a satellite at higher altitudes would produce appreciable torques as a result of solar radiation pressure. This problem can be solved, however, by deploying an additional boom or booms out from the center of gravity of the satellite so that the solar torques are balanced.

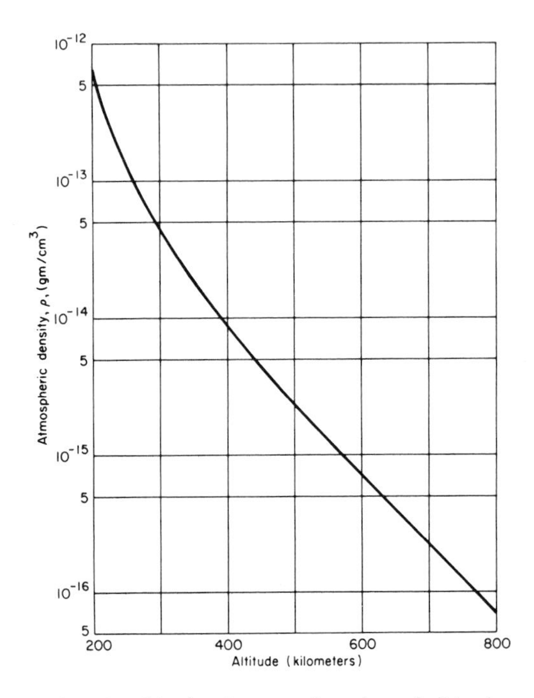

FIG. 8. Air density as a function of altitude.

Aerodynamic drag can disturb the vertical orientation for satellites at low orbital altitudes. H. K. Kallmann [12] has obtained a curve (Fig. 8) giving the atmospheric density as a function of altitude. It can be readily shown that the drag force on an orbiting body is given by [13]

$$\tau_A = \tfrac{1}{2} C_d A_p \rho v^2 d \qquad \text{(dynes)} \tag{25}$$

where

C_d = ballistic coefficient

ρ = atmospheric density at satellite altitude (gm/cm^3)

v = orbital velocity (cm/sec)

For the Traac satellite configuration as shown in Fig. 7, the aerodynamic torque for an altitude of 500 miles would be 67 dyne-cm, resulting in an approximate deviation off the vertical of 0.1 deg. At an altitude of 300 miles the displacement off the vertical would be approximately 5 deg. It is possible to solve the aerodynamic torque problem for satellites at very low altitudes by the use of additional booms that balance the torque about the satellite's center of gravity.

An additional perturbation to vertical stabilization will result when the gravity-gradient stabilized satellite is in an eccentric orbit. Dr. Vanderslice [14] of the Applied Physics Laboratory has shown that, for a well-damped satellite as described above, the maximum deviation off the vertical is given by the eccentricity expressed in radians. For example, an eccentricity of 0.02 would cause a maximum deviation off the vertical of 0.115 deg.

VI. Conclusions

A satellite can be designed to passively stabilize along the direction of the earth's gravity gradient. A coated spring in conjunction with hysteresis damping rods can be used to effectively damp the librations of the satellite about the local vertical direction. A spring designed for this purpose was successfully deployed on the Traac satellite and did damp oscillations about the local magnetic field direction. The satellite can be designed so that the perturbing torques are quite negligible in comparison to the gravity-gradient torque.

References

1. R. E. Fischell, Magnetic and Gravity Attitude Stabilization of Earth Satellites, *Johns Hopkins Univ., Appl. Phys. Lab. Rept. CM-996,* May 1961.
2. R. E. Roberson, Gravitational torque on a satellite vehicle. *J. Franklin Inst.* **265**, 13-22 (1958).
3. W. B. Klemperer and R. N. Baker, Satellite librations. *Astronaut. Acta* **3**, 16-27 (1957).
4. R. A. Nidley, Gravitational torque on a satellite of arbitrary shape. *ARS J.* **30**, 203-204 (1960).
5. R. B. Kershner, and R. R. Newton, Attitude control of artificial satellites. *In* "Space Astrophysics," Chapter 14. McGraw-Hill, New York, 1961.
6. R. E. Fischell, Magnetic damping of the angular motions of earth satellites, *ARS J.* **31**, 1210-1217 (1961).
7. R. E. Fischell, Passive magnetic attitude control for earth satellites. *8th Ann. Meeting Am. Astronaut. Soc., Washington, D. C., January 16, 1962,* Paper 62-8.

8. L. J. Kamm, "Vertistat"—An improved satellite orientation device. *ARS J.* **32**, 911-913 (1962).
9. Damping spring for gravity-stabilized satellites. *APL Tech. Dig.* **2**, No. 2, 20-21 (1962).
10. R. E. Fischell, The Traac satellite. *APL Tech. Dig.* **1**, No. 3, 2-9 (1962).
11. B. Paul, Planar librations of an extensible dumbbell satellite. *AIAA J.* **1**, 411-418 (1963).
12. H. K. Kallmann, A preliminary model atmosphere based on rocket and satellite data. *J. Geophys. Res.* **64**, No. 6, 615-623 (1959).
13. F. F. Mobley, private communication. Johns Hopkins Univ. Appl. Phys. Lab.
14. J. L. Vanderslice, Dynamic analysis of gravity-gradient satellite with passive damping. *Johns Hopkins Univ., Appl. Phys. Lab. Rept. TG-502*, June 1963.

The Application of Gyrostabilizers to Orbiting Vehicles

J. E. DeLisle, E. G. Ogletree, and B. M. Hildebrant

Instrumentation Laboratory, Massachusetts Institute of Technology, Cambridge, Massachusetts

I. Introduction

A. General

Among the first references in the unclassified literature to the use of gyrostabilizers for satellite attitude control is an article by Roberson [1] in 1957. In another article [2] by the same author, reference is made to unpublished work by D. L. Freebairn and others at Autonetics in 1953. Further theoretical investigation of the use of gyrostabilizers in orbiting vehicles was done at the Massachusetts Institute of Technology Instrumentation Laboratory in 1959 [3], and by Burt in 1961 [4].

A properly designed gyrostabilizer system can function not only as a damper, but also as a means of satellite attitude control. Yaw and roll can be stiffened to reduce the offsetting effects of steady torques, while all axes can be made resistant (to a greater or lesser degree) to dynamic torques of an impulsive or time-varying nature.

Since the use of a gyrostabilizer as a semipassive damping device for orbiting satellite vehicles appeared so attractive from the standpoint of simplicity, reliability, and effectiveness, work at the Instrumentation Laboratory on the development of gyrostabilizer systems has continued. In addition to the work reported in reference [3], further analysis and study of the problem has appeared in later Massachusetts Institute of Technology reports [5-8]. It is these later reports which form the basis for most of this review.

B. Definition of a Gyrostabilizer

A gyrostabilizer is considered to be a rotating wheel gimbaled in such a way that the spin axis of the wheel, which defines a characteristic angular momentum vector, can precess about an axis (output axis)

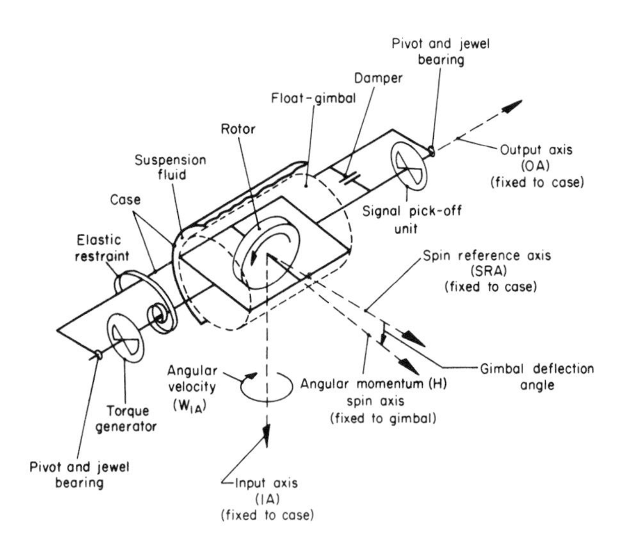

FIG. 1. Diagram of a single-axis floated gyro.

normal to the spin axis in response to angular rates about a third axis (input axis) mutually perpendicular to both the spin and input axes. In short, a gyrostabilizer is a gyroscope (or gyro) as depicted in Fig. 1, but used in such a way that it not only senses rates but supplies torques directly to the vehicle to be stabilized. Damping of gimbal precession rates will be present, as well as spring restraints about the output or precession axis. The signal generator and torque generator may be used for signal monitoring and/or for feedback or control purposes to modify the system response. For the class of gyrostabilizer systems considered here, a torque generator or some other form of torquing device will generally be required to apply torques about the output axis for those applications where the nominal angular momentum vector (spin reference axis, SRA) of the gyrostabilizer does not lie along the orbital rate vector.

C. Types of Vehicles to Which Gyrostabilizers Are Applicable

Gyrostabilizers may be practically applied to many types of vehicles, including interplanetary or orbiting vehicles (including those that are spin stabilized). However, this discussion will be concerned with earth-oriented satellites in nearly circular orbits. The earth orientation requirement probably restricts the application to vehicles of synchronous orbit and lower as a practical matter, since unbalanced solar radiation pressure may have upsetting effects comparable to the orienting effect of the gradient of the gravitational field for higher orbits.

It is believed that gyrostabilizers in conjunction with earth-oriented vehicles offer many attractions for satellites used for: (1) communication, (2) weather surveillance and reconnaissance, and (3) space experiments. In general, the applicability of this stabilizing means is to reliable, modest-lifetime—6 months to several years—vertically oriented satellites where an accuracy of $\frac{1}{2}$ to 5 deg in all axes is desired. Indeed it has also been shown [9, 10] that high pointing accuracy can be achieved if external sensors are used to command the torque generators about the precession axes. In such applications, the automatic sensing and exertion of appropriate stabilizing torques would normally be retained.

II. The Satellite Damping and Stabilization Problem

A. General

Before proceeding further with the discussion of the mechanics of gyrostabilizers, it might be well to examine and define the problems that one wishes to deal with. To state this in a general way is not too difficult.

A satellite is to be injected into a nearly circular orbit at some prescribed altitude. It is to settle rather quickly, with the axis of least inertia along the vertical, and the axis of maximum inertia normal to the orbital plane, and to hold this position in the face of torque disturbances. Perhaps at this point it can be seen that a fast settling time may not be compatible with a system that remains unperturbed in the presence of large disturbances, since settling is brought about by the relatively modest gravitational-gradient torque on the vehicle. There are other effects, such as eccentricity, that may also require study to see what happens to impulse response, say, when eccentricity response is minimized.

B. Transient Damping

The gyrostabilizer system is expected to function as a transient damping device since there will generally be satellite position and rate errors when it is first injected into orbit, i.e., the satellite will not be perfectly aligned to the vertical initially. A reasonable settling time will be desired. There may be disturbances (e.g., due to micrometeorite impacts) that cause vehicle offsets, in which case a similar settling time will again be required for the gravitational-gradient torque to bring the vehicle to the desired position.

C. Stability Augmentation

As a body possessing inertia, the satellite has a certain calculable impedance to external and internal torques, both impulsive and continuous, without the addition of a stabilizing or damping means. The gradient of the gravitational field provides a restoring torque in roll and in pitch against steady disturbing torques about these axes. The roll axis receives an additional steady restoring torque owing to gyroscopic effects of the satellite as a rotating body, while the yaw axis receives virtually its only restoring torque as a result of gyroscopic effects. The gravitational-gradient torque will be discussed in more detail in subsequent sections.

III. Coordinate Reference Frames

In the preceding sections, the problems involved in stabilizing an earth-oriented satellite were introduced, and the nature of a single-axis gyroscope employed as a gyrostabilizer was described. In order to be more explicit regarding the design of gyrostabilizer types of systems, it will be useful to first define an adequate set of coordinate reference frames.

Although many types of reference frames are available, the right-handed orthogonal Cartesian coordinate system has been chosen for use throughout this report. The simplicity and directness of analysis based on such frames is helpful in retaining the over-all goals of an analytic manipulation. It is suggested that much insight into the three-dimensional multivariable problem of satellite dynamic motion may be gained by a careful study of the reference frames illustrated here, namely:

Geocentric inertial reference frame	I	(Fig. 2)
Geocentric earth reference frame	E	(Fig. 2)
Geocentric orbital plane reference frame	K	(Fig. 3)
Geocentric orbital position reference frame	P	(Fig. 3)
Vehicle-centered orbital reference frame	O	(Fig. 4)
Vehicle-centered vehicle reference frame	V	(Fig. 5)
Vehicle-centered principal axis reference frame	A	(Fig. 6)
Vehicle-centered gyro case axis reference frame	GU	(Fig. 7)
Vehicle-centered gyro gimbal axis reference frame	GIM	(Fig. 8)

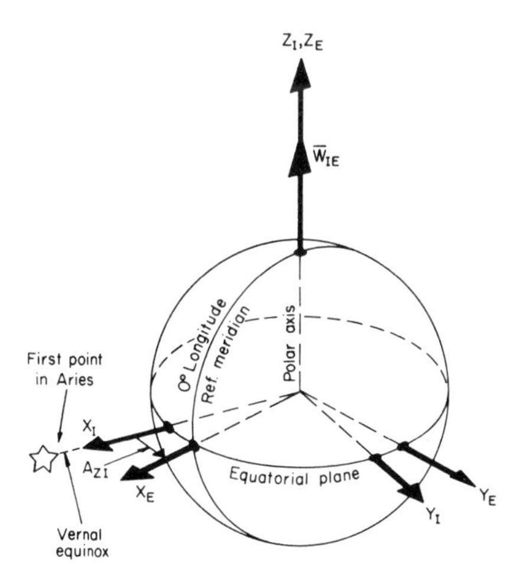

FIG. 2. Geocentric inertial (I) and earth (E) reference coordinate frames. (Note that vectors are identified by the vinculum in all the figures.)

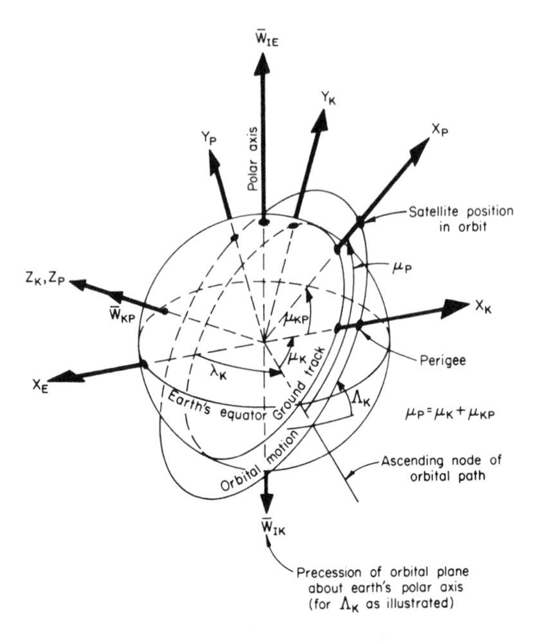

FIG. 3. Geocentric orbital plane (K) and orbital position (P) reference coordinate frames.

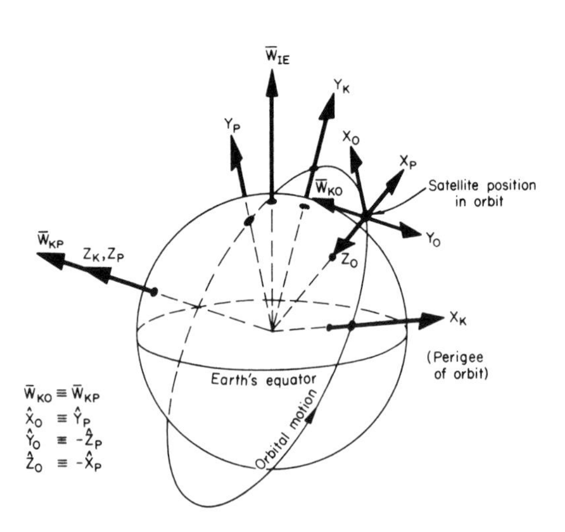

FIG. 4. Vehicle-centered orbital (O) reference coordinate frame.

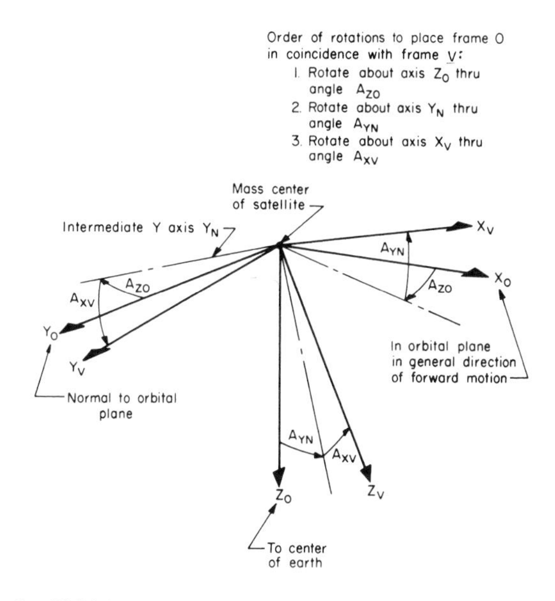

FIG. 5. Vehicle-centered vehicle (V) reference coordinate frame.

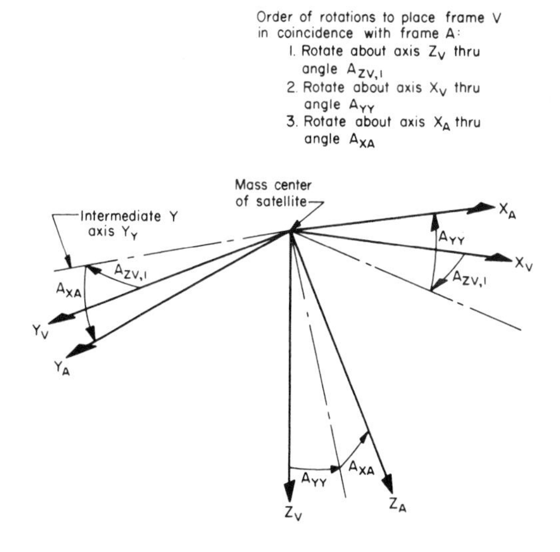

FIG. 6. Vehicle-centered principal axis (A) reference coordinate frame.

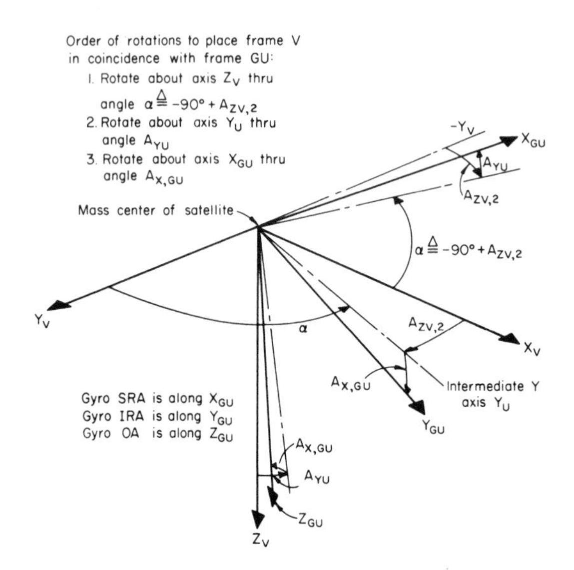

FIG. 7. Vehicle-centered gyro case axis (GU) reference coordinate frame.

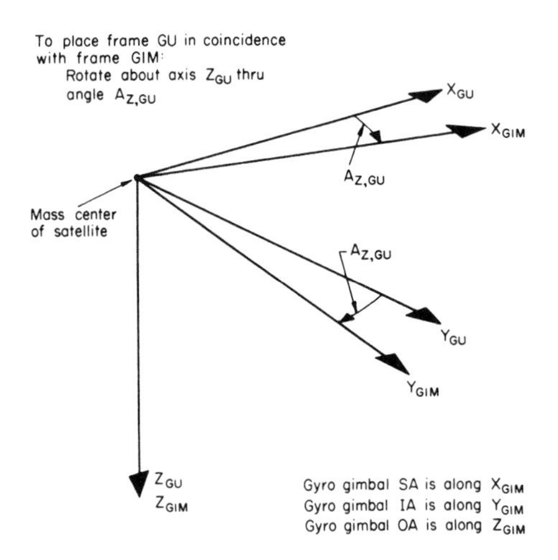

FIG. 8. Vehicle-centered gyro gimbal axis (GIM) reference coordinate frame.

Comments on these frames may be helpful at this point. The gimbal of a particular gyrostabilizer in a system has a particular GIM frame associated with it. It has a single rotational degree of freedom relative to its associated gyro case, which, in turn, defines a particular GU frame. The gyro case, being rigidly attached to the vehicle structure, has no rotational freedom relative to frame V which is defined typically by a set of lines inscribed in the rigid structure of the satellite. An assumption, usually valid for orbiting vehicles, is that the vehicle is designed to ensure that the principal axes are invariant within the vehicle structure; thus, frame A is assumed to be at rest relative to frame V. (In the present paper, it will be assumed that frames V and A are coincident except as otherwise noted under the specialized discussions of skewed vehicle configurations.)

In all of the cases considered here the desired objective of the stabilization and control system is to cause the V frame to instrument (i.e., track, tend toward coincidence with, etc.) the orbital reference frame O. As shown in Fig. 4, a vehicle that is properly stabilized will then indicate the local vertical by its Z_V axis, the normal to the orbital plane by its Y_V axis, and—in the case of a circular orbit at least—the forward direction of flight by its X_V axis.

The orbital plane, and the geocentric radius to the satellite as it moves in the plane, are defined by the K and P frames, respectively. Actually, planar motion of the satellite mass center about the earth is not possible without the application of small corrective accelerations normal to the orbital path, due to the irregularities in the earth's gravitational potential field. (This point is treated at length in the recent literature of orbital mechanics. An introductory discourse by Dr. Roberson is included in his section on Attitude Control in reference [11]). However, the orbital motion is essentially planar for most missions and the K frame, having some inertial rotation $\mathbf{W}_{IK}$ (which may be zero), provides an adequate analytic representation of an orbital "plane," particularly for introductory papers such as the present one.

The E frame, imbedded in the rotating earth, and the I frame, affixed to a set of inertially nonrotating star lines, are familiar to most readers in this field and need no further introduction here. For example, see reference [12].

These and other pertinent frames are presented and discussed in detail in Appendix A of references [5, 6]. Matrix representations of the orthogonal coordinate transformations linking at least adjacent frames are included therein, and a notation is presented that facilitates derivation of the equations of rotational motion of the satellite and its stabilization and control system in a reasonably simple, direct, and orderly manner. A somewhat more concise development, specialized to a particular gyrostabilizer system (the "vertical-vee" or "roll-vee" configuration to be described later in this paper) is to be found in Section II of reference [8].

IV. The Satellite as a Gyro Element

The next step in the process of developing a conceptual understanding of gyrostabilizer systems is to visualize the inherently gyroscopic nature of satellite rotational dynamics. In the case of rapidly spinning satellites such as the Telstar, Tiros, Transit, and similar vehicles, the gyroscopic nature of the body is readily apparent. On the other hand, for earth-oriented vehicles such as those considered here, recognition of the satellite as a gyro element is somewhat more difficult. This is particularly true, for example, of a synchronous equatorial satellite stabilized to the O frame in the manner indicated above. Such a vehicle, nonrotating as viewed from the earth, would seem to an earth-fixed observer to have little resemblance to a gyro element. More about this will follow, but first a word about the net effect on the satellites studied here of the gradient of the earth's gravitational field.

In explaining the gravitational-gradient torque on a satellite it is customary to depict the vehicle as a dumbbell consisting of two point masses separated by a massless rod. Consider that the rod is deflected from parallelism with the geocentric radius to the mass center of the satellite. The gravitational force on each point mass, in the direction of the earth's center, is given by:

$$F_L = \frac{Em}{R_L^2} \qquad \text{and} \qquad F_U = \frac{Em}{R_U^2} \tag{1}$$

where the subscripts L and U refer to the lower and upper mass (each of mass m) respectively, E is the gravitational constant of the earth, and R_L and R_U are the geocentric radii to the lower and upper masses respectively. Since R_U is greater than R_L it is evident that F_L exceeds F_U. With a misaligned dumbbell, neither of these forces will act through the mass center of the dumbbell. Hence, each will exert a turning moment on the dumbbell about the mass center. Since the moment produced by the force on the lower mass exceeds that produced by the force on the upper mass, the net effect will be a torque that produces the rotation of the dumbbell axis toward parallelism with the local vertical through the smaller of the two possible angles. This torque varies as three-halves of the sine of twice the angle of deflection of the axis of least inertia from the geocentric radius about any horizontal axis, and has received extensive treatment in the literature (e.g., references [5, 13]). It will be evident from subsequent remarks that it provides the only restoring moment influencing pitch axis motions of the satellites studied here and contributes to the total roll axis restoring torque. However, it has no component about the orbital yaw axis, since the opposing forces described above can have no moment *about* the geocentric radius.

Returning now to the visualization of the vehicle as a gyro element, it should be noted that the term *gyro element* as used here refers to a body having finite, nonzero mass, and which is characterized by a nonzero inertial angular velocity about some axis through its mass center.[1] The motions of such a body are governed by the law of conservation of angular momentum [14-16]. Specifically, if, as seen by an observer fixed in the I frame, there is no unbalanced torque acting on the gyro element, then that same observer will see the total angular momentum of the gyro element as a vector that is invariant both in magnitude and in direction. Conversely, if the observer sees a vector unbalanced torque acting on the gyro element, he will simultaneously observe, as an identical vector quantity, a nonzero time rate of change of the angular momentum of the element. These concepts are neatly condensed into the rotational form of Newton's second law of motion:

$$\left.\frac{d\mathbf{H}}{dt}\right|_{I} = \sum \mathbf{M}\Big|_{\text{applied}} \tag{2}$$

That is, the rate of change of total angular momentum of the rotating body (including that of any internal parts) with respect to time, as seen in inertial space, is equal to the vector sum of all externally applied torques. Equation (2) is the fundamental equation of gyroscopic analysis. It is therefore the fundamental equation of satellite motion analysis, particularly for the type of satellite studied here, since such satellites are gyro elements. This may be visualized as follows.

For simplicity, assume a circular orbit about a spherically symmetric earth. The orbital motion of the mass center of the satellite may then be ignored temporarily while we try to achieve a clear understanding of the rotational dynamics about its mass center. An observer placed at the satellite mass center, nonrotating with respect to the orbital reference frame O, and looking downward toward the center of the earth, would consider himself to be inertially fixed above a rotating earth. He would also be aware of the libratory motion of the satellite as frame V—affixed to the vehicle structure—exhibited oscillations about its equilibrium orientation in which it tracks frame O. Such an observer would also be able to predict certain of the external torques acting on the satellite. For example, there is the gravitational-gradient torque about a horizontal

[1] Actually, satellites of this type have certain of the dynamic characteristics of the gyro element as more rigorously defined in derivation summary no. 1 of reference [16].

axis through the mass center, produced by misalignment of the axis of least inertia (the Z_A axis) with the local vertical (the Z_O axis).

The O frame observer might reason that, since misalignment (yaw) of the vehicle *about* the Z_O axis (the local vertical) gives rise to *no* external gravitational-gradient torque (i.e., the spherical earth model is just as spherical, viewed from any yaw aspect), the orientation of the vehicle about the local vertical is arbitrary. However, the vehicle, given such an initial misalignment in yaw and released, would exhibit not only oscillations about the local vertical but also coupled oscillations about the horizontal axis in the orbital plane (the X_O axis).

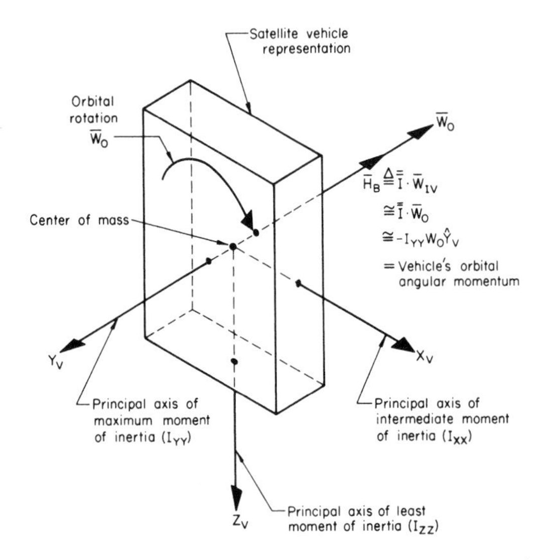

FIG. 9. Representation of the satellite vehicle as a gyro element.

In order to justify the observed phenomena in the case of the undamped vehicle alone, the O frame observer might postulate an inertial angular velocity about the negative Y_O axis and therefore a vector angular momentum associated with the vehicle itself and directed generally along the negative sense of the Y_V axis. Such an angular momentum vector *is* in fact associated with the vehicle owing to its once-per-orbit revolution about its mass center in pointing nominally "downward" toward the earth's center at every point along its path. The vehicle's angular momentum $\mathbf{H}_B$ is illustrated in Fig. 9. The difficulty in presenting this point is in finding a single observer, associated with the satellite's position, that can be aware of both the rotational and libratory motions of the satellite. Perhaps a better choice would be an observer that is nonrotating with respect to frame K (see Fig. 3) but is translating along with the mass center of the satellite.

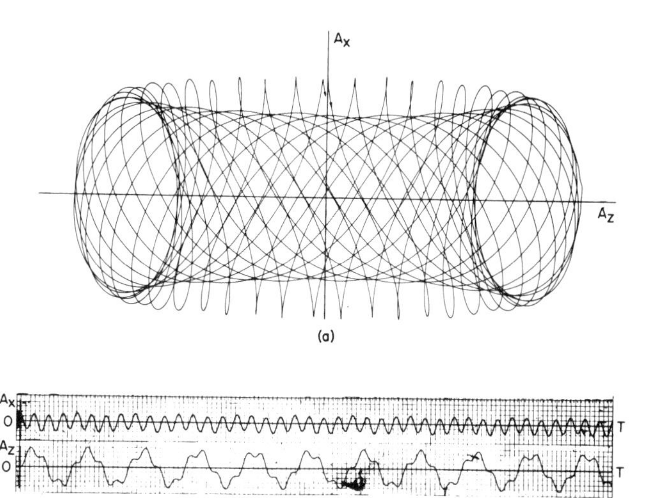

FIG. 10. (a) Example of roll vs. yaw error angles for an undamped satellite with initial roll error angle. (b) Roll and yaw error angles vs. time for the curve of (a).

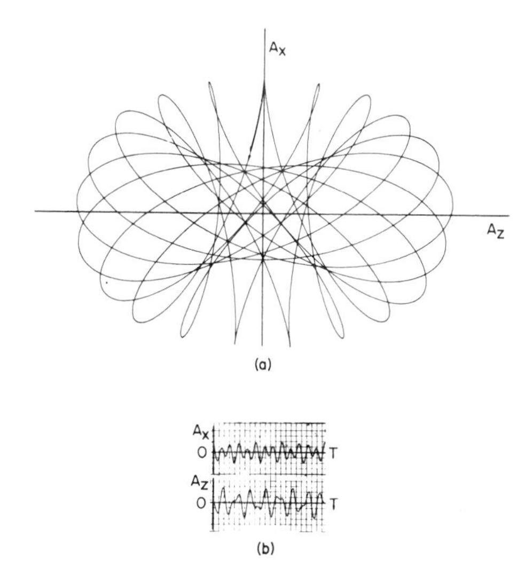

FIG. 11. (a) Example of roll vs. yaw error angles for an undamped satellite with constant speed wheel along negative pitch axis, for initial roll error angle. (b) Roll and yaw error angles vs. time for the curve of (a).

The simultaneous nature of roll and yaw motions may be clarified by plotting them as functions of each other, with time suppressed as a coordinate. If the roll error angle[2] A_X is plotted as a function of the yaw error angle A_Z for an initial roll offset, the result for an undamped vehicle is ordinarily a complex curve that is described within a smooth envelope that is nominally centered about the origin in the absence of constant external torques. The shape of the curve is dependent upon the satellite vehicle configuration and the resultant angular momentum of any rotating components within the vehicle. Figure 10 is a representative plot for a particular undamped vehicle with no internal rotating parts; while Fig. 11 shows the effect on the plot if the same vehicle is equipped with a wheel spinning with a particular value of constant angular momentum about the negative Y_V axis. (It should again be noted that the shapes of these plots are a definite function of the vehicle configuration, the angular momentum of the wheel, and the initial conditions chosen; they are included here merely to illustrate the coupled nature of roll and yaw motions.)

It is evident from the above that the satellite vehicle of this discussion has certain of the characteristics of a gyro element. If external torques are restricted to the plane normal to the angular momentum vector, the gyro equation reduces to:

$$\mathbf{W}_P \times \mathbf{H} = \sum \mathbf{M} \Big|_{\text{applied}} \tag{3}$$

where $\mathbf{W}_P$ is the vector precessional angular velocity of the angular momentum vector $\mathbf{H}$ in response to the externally applied torques. Since its angular momentum is nominally along the negative sense of the pitch (Y_V) axis, the vehicle will respond to unbalanced roll axis torques with precession about the yaw axis and to unbalanced yaw axis torques with precession about the negative sense of the roll axis. Unbalanced pitch axis torques will not produce gyroscopic precession, but will only alter the magnitude of the body's angular momentum vector. The pitch axis error angle for a particular undamped vehicle, plotted as a function of time after an initial pitch offset, is as shown in Fig. 12.

From a consideration of the error plots described above, it is evident that the vehicle axes may be caused to settle to an equilibrium condition in which they track the orbital reference axes, frame O, if damping torques are applied both about the pitch axis and in the roll-yaw plane.

[2] The Euler error angles A_X and A_Z used here are the angles designated as A_{XV} and A_{ZO}, respectively, in Fig. 5. Similarly, the pitch error A_Y, to be introduced later, corresponds to the angle A_{YN} in the same figure.

As in gyrocompass erection, the satellite roll-yaw damping may be accomplished by damping the motion about either the roll or the yaw axis or both, since motions about the two axes are gyroscopically coupled. The conventional settling of pitch error with the application of pitch axis damping is shown in Figs. 13a and 13b for relatively light and relatively heavy damping, respectively. The settling of roll and yaw motions in the presence of damping, for a particular vehicle and damping system, might be as illustrated as in Figs. 14 and 15 for two different sets of initial conditions.

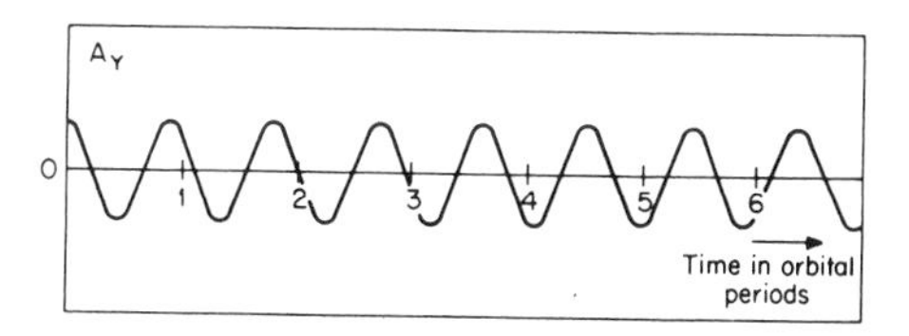

FIG. 12. Pitch error angle vs. time for an undamped satellite with initial pitch error.

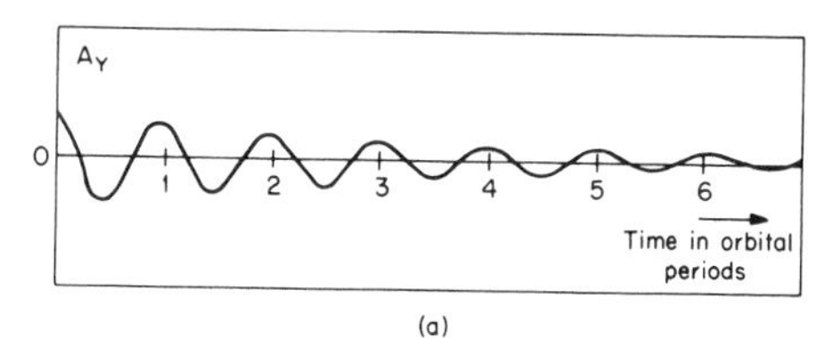

(a)

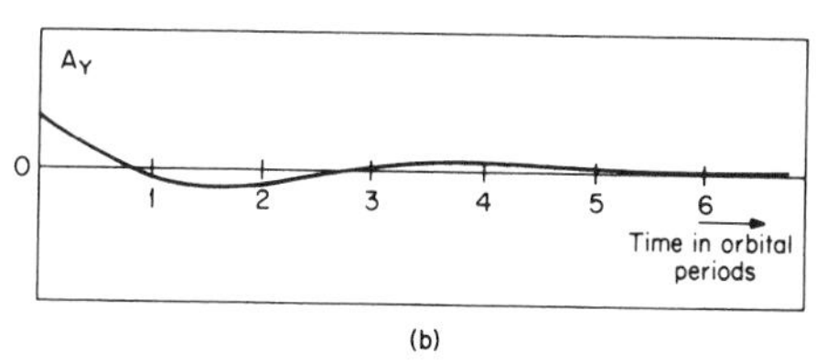

(b)

FIG. 13. (a) Pitch error angle vs. time for relatively light damping with initial pitch error. (b) Pitch angle error vs. time for relatively heavy damping with initial pitch error.

The above remarks may be summarized as follows: A satellite of this type possesses certain of the dynamic characteristics of a gyro element; with suitable damping of its motions, axes fixed in such a vehicle can be made to instrument a local orbital reference frame; the accomplishment of the desired damping involves two separate damping torque effects, namely, one along the axis that is to be normal to the orbital plane and one that is along any direction in the orbital plane.

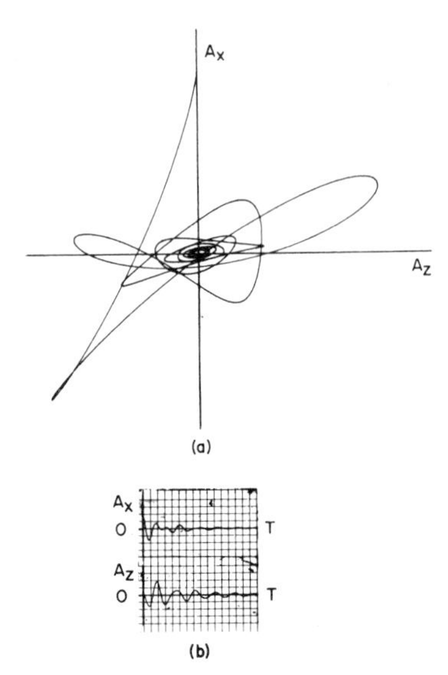

FIG. 14. (a) Example of roll vs. yaw error angles for a satellite having a roll-yaw damping system, for initial roll error angle. (b) Roll and yaw error angles vs. time for the curve of (a).

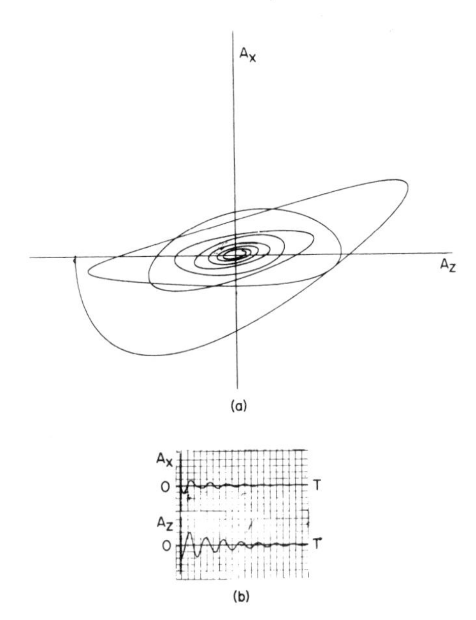

FIG. 15. (a) Example of roll vs. yaw error angles for a satellite having a roll-yaw damping system for negative initial steps in yaw error angle and roll error angle rate. (b) Roll and yaw error angles vs. time for the curve of (a).

V. Attitude Control Systems Using Gyrostabilizers

Vehicle pitch axis motion, for the satellite discussed above, may be damped by the application of pitch axis torques proportional to pitch error rates. Techniques for accomplishing this by means of gas jets, inertia reaction wheels, passive wheels and fluids with viscous pitch axis coupling, on-board electromagnets, and gyroscopes have been explored at some length in the literature [2, 5, 17-20]. The emphasis in the present paper is on semipassive gyroscopic (i.e., gyrostabilizer) methods of achieving pitch axis control. The accomplishment of such pitch axis damping, using multiple gyro configurations, has been described in reference [5].

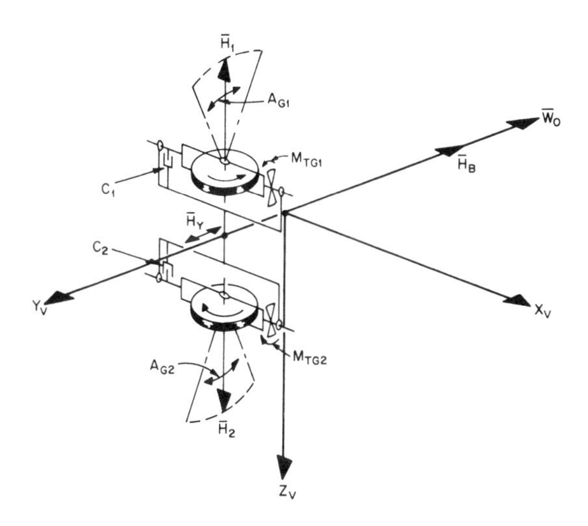

FIG. 16. Pitch axis damping system using two gyrostabilizers in a "back-to-back" configuration.

A mechanization for pitch axis damping with a "back-to-back" pair of integrating gyros is illustrated in Fig. 16. The gyros are arranged to have no component of angular momentum along the roll axis, and for their components of angular momentum along the yaw axis to be continuously equal and opposite in order to be self-canceling. Thus, the system has no influence on the roll-yaw motion. As the gyros precess equally and oppositely in response to angular velocities occurring along their common input reference axis—which is the vehicle pitch axis—the vector sum of the two spin angular momentums, $\mathbf{H}_1$ and $\mathbf{H}_2$, is a pitch axis angular momentum component, $\mathbf{H}_Y$. Since the gyro gimbal precession is restrained by the viscous coupling (C) in each gyro and the instruments hence are classed as integrating gyros, it follows that each gimbal's angular velocity—and hence the time rate of change of the pitch axis component of total gyro angular momentum $\dot{H}_Y$—is proportional to the sensed input axis angular velocity. The vehicle's pitch axis angular velocity is composed of orbital rate (a constant for the circular orbit studied here) and pitch error rate. The constant component would produce a continuous gimbal precession if uncompensated. It must be canceled in each gyro by the application of constant torque generator torques M_{TG1} and M_{TG2} leaving only the pitch error rate, $\dot{A}_Y$, influencing gimbal precession. Thus, the rate of change of damping system angular momentum along the pitch axis (equivalent to an applied torque about the pitch axis) is directly proportional to pitch error angular velocity, and therefore constitutes a damping torque. Suitable selection of gyro angular momentum and coefficient of damping will yield any desired degree of vehicle pitch axis damping. Note that the system would perform in exactly the same manner if the roll and yaw axes in the figure were interchanged and phased properly to retain a right-handed set.

It should also be noted that in the pitch axis damping system just described there is no requirement for separate sensing, computing, and control devices. The integrating gyroscopes provide all three functions. This is a principal characteristic of gyrostabilizers as a class. It is only necessary to provide the correct orientation of the gyros in the vehicle, the power needed to spin their wheels at constant speed, and the steady gimbal torques needed to prevent gyroscopic precession in response to the body's orbital angular velocity about its mass center. The pair of gyros then serves as a self-contained pitch damping system. Hence, the term "semipassive" is used to describe these systems. This simplicity is characteristic of the gyroscopic control systems discussed here. None of them require any external error sensing by means involving electromagnetic radiation, nor is any on-board computation, beyond that inherent in the gyroscopic responses, involved.

The use of two gyros in a back-to-back arrangement as described above is applicable to damping of motion about any axis, not just pitch. Special torques would have to be applied in each case via the torque generators (or similar means, such as spring restraints) to keep the gyro angular momentums nominally zero in the plane normal to the controlled axis, but the damping function will be essentially the same along the roll or yaw axis, for example, as it is along pitch. The actual response in roll or yaw is complicated, of course, by the gyroscopic coupling of the two axes, but the resulting torques due to precession of the gyro gimbals are, in fact, damping torques, and two-axis settling is obtained. Somewhat more damping is attainable by using a pair of gyros in this manner along *each* of the three axes owing to the larger number of adjustable parameters, but the single pair with input axes in the roll-yaw plane ordinarily provides good settling if properly designed. It will be shown in the following discussion that there are several simpler mechanizations offering roll-yaw as well as three-axis damping.

Let us return now to the vehicle of Fig. 9 and recall that the body, in having in its equilibrium orientation an angular momentum normal to the orbital plane, has in effect an infinitude of gyroscopic input axes that form the orbital plane. We next consider simpler methods of achieving damping in roll and in yaw with gyrostabilizers. Figures 17 and 18 show

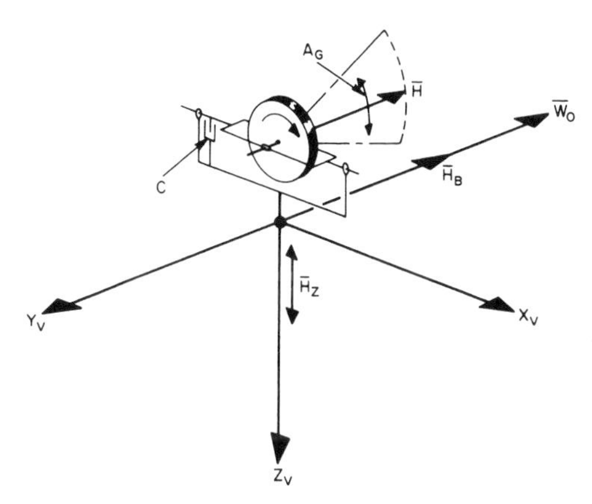

FIG. 17. Roll-yaw damping system using a single gyrostabilizer with input axis along yaw axis.

two roll-yaw damping systems, each of which involves only a single gyrostabilizer. In both cases, the single gyro's angular momentum **H** is nominally aligned with the body's angular velocity vector $\mathbf{W}_0$ in order not to exert unwanted body torques in the steady state due to gyroscopic coupling with $\mathbf{W}_0$. The input axis in Fig. 17 is in the yaw axis direction, while that in Fig. 18 is along the roll axis. The action of the gyro in either case is such that its precession in response to error angular velocities about the input axis produces damping torques about that axis. As shown previously in Figs. 14 and 15, this results in both roll and yaw settling. Experience has shown that there is an upper bound to the combined roll and yaw axis damping that can be obtained with this sort of system. The optimum response, giving equal weight to settling about both axes, generally occurs if the angular momentum of the single gyrostabilizer is of the same order of magnitude as the orbital angular momentum $\mathbf{H}_B$ of the vehicle itself [4, 5]. This fact emphasizes the type of roll and yaw damping action that occurs in semipassive systems utilizing gyrostabilizers. It is this: The body's motion is damped by the relative precession of the gyro gimbal or, alternatively, the gyro's gimbal motion is damped by the relative precession of the body. The one "rubs against" the other, and, perhaps in some way analogous to the principle of maximum power transfer in electrical circuits, the maximum rate of energy dissipation in the gyro's damper seems to occur when the two rubbing parts (vehicle and gimbal) are about equal in angular momentum.

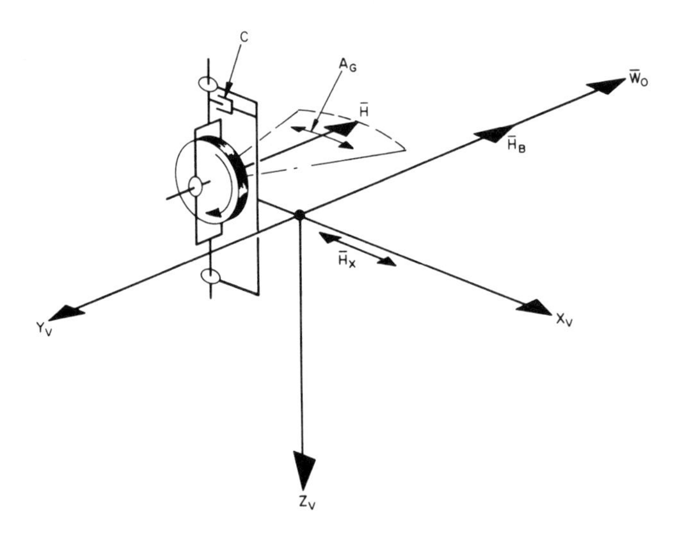

FIG. 18. Roll-yaw damping system using a single gyrostabilizer with input axis along roll axis.

Since the gyro angular momentums of Figs. 17 and 18 are approximately determined by vehicle inertias and orbital angular velocity, the principal parameter for optimizing system response is the H/C ratio.[3] (Actually, the torque constant of flex leads, other springs, or electronic feedback loops around the gyros gives additional parameters, but generally the performance is degraded for positive spring constants, and the stability of the system is to some extent reduced in the case of negative spring constants.) By combining the concepts of Figs. 16 and 17, it is possible to achieve three-axis control with only two gyros, and to have

[3] Certain specifications may dictate other values of gyro angular momentum. However, when transient response is the chief criterion, the above rule of thumb is valid.

three adjustable parameters, within certain physical constraints. Such a system is illustrated in Fig. 19 and will be referred to here as a "roll-vee" or "vertical-vee" configuration. The components of angular momentum of the two gyros along the yaw axis comprise the pitch damping system analogous to that of Fig. 16. The components of gyro angular momentum along the negative pitch axis provide roll-yaw damping in the same manner as the system of Fig. 17. Note that the gyros could have been arranged with their output axes along yaw rather than roll. This would lead to a system described here as a "yaw-vee" configuration. The equations of motion with either of the two vee configurations, as well as the vehicle and control system responses to roll or yaw disturbances, are quite similar. The responses to pitch disturbances are identical.

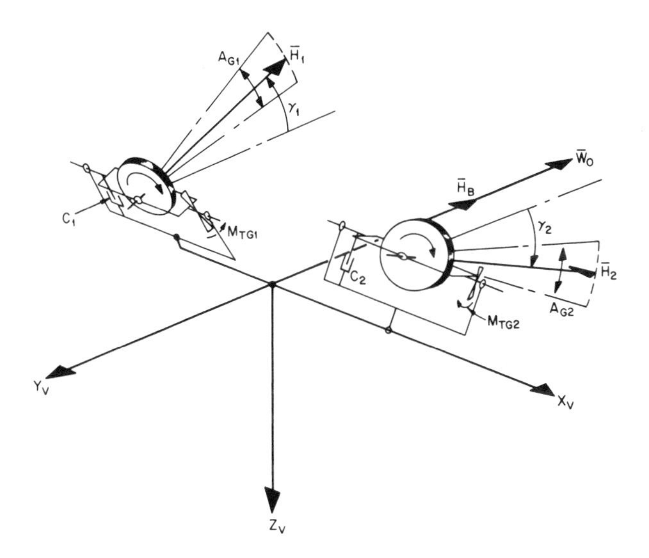

FIG. 19. Three-axis damping system using a roll-vee configuration of two gyrostabilizers.

The choice of skew angle γ for a particular size of gyro angular momentum H in the two vee configurations seems to follow the conclusions presented in connection with Figs. 17 and 18. That is, the sum of the projections of the gyro angular momentum vectors on the negative pitch axis ($2H \cos \gamma$) should be of the order of magnitude of the vehicle's orbital angular momentum for optimum roll-yaw response. Obviously, as larger skew angles are employed, with attendant increase in gyro angular momentum (to keep $2H \cos \gamma$ near optimum), the pitch axis damping increases. Burt [4] has pointed out that, in this sort of system, the optimum pitch transient response occurs with the system adjusted to provide three equal, real, negative roots in the third-order characteristic equation of pitch motion. On the other hand, again referring to Burt's results, the pitch error response to the forcing effects of eccentricity may be greatly reduced by severely over-damping the pitch system, but this compromises its transient response, as expected. This subject will be discussed in more detail in a later section on optimization.

Thus, the vee configurations appear to offer possibilities of quite flexible three-axis damping, combining the capabilities of the three-gyro systems represented by Figs. 16 and 17 or 16 and 18. As mentioned previously, there are bounds on the attainable roll and yaw axis damping, since the vehicle is called upon as a gyro element itself to work against the gyro angular momentum components that comprise the roll-yaw damping system. If negative spring constants are allowed between each gyro gimbal and its respective case, and a spring and dashpot arrangement (with negative spring constant allowed) linking the two gimbals directly is employed, Burt [4] has shown that remarkably good three-axis response is attainable. Settling times (to $1/e$ of small initial steps in error angles) of the order of fractions of an orbital period are possible with certain vehicle and system configurations. The effects of flex lead or other elastic torques on the gimbals must be included in any analysis or simulations of these systems, as they may have a profound influence on the results as compared with those obtained when the gyrostabilizers are created as pure integrators.

At this point it has been demonstrated that the satellite vehicle itself behaves as a gyro element (Fig. 9), and can be made to "work against" a conventional integrating gyro in damping roll and yaw motions, as in Figs. 17 and 18. Also, when two gyros are arranged in a vee configuration as in Fig. 19, the combination will accomplish three-axis vehicle damping. We might examine what would happen to the satellite equipped with such a vee configuration of two gyrostabilizers if one of the gyros were removed, leaving only a single gyro skewed about the vehicle's roll or yaw axis. The angular momentum of such a single skewed gyrostabilizer can be resolved into a component along the negative pitch axis and (taking the yaw-skewed case as an example) the roll axis. Clearly, the component along the roll axis will couple with the orbital angular velocity $\mathbf{W}_0$ and, because of the presence of the torque generator, will result in the exertion of a gyroscopic torque in yaw on the vehicle. Such an unbalanced yaw torque will deflect the vehicle about the yaw axis. An equilibrium condition will result (assume for the moment that the response to the unbalanced torque has been damped, and that the torque generator torque has been reduced if required for maintenance of system stability), in which the body's own gyroscopic restoring torque in yaw is equal and opposite to the steady state torque exerted on the vehicle by the single gyrostabilizer via its torque generator. What has happened here is that the body's orbital angular momentum vector has been deflected from its customary alignment with $\mathbf{W}_0$ and caused to be skewed about the yaw axis such that the cross-product of body angular momentum with $\mathbf{W}_0$ is a vector torque equal in magnitude and opposite in sense to the cross-product of gyrostabilizer angular momentum and $\mathbf{W}_0$. (The magnitude of these torques is the new constant torque level that must be exerted by the gyro's torque generator. It will generally be a lower level than was needed in the vee configuration since the body's yaw restoring torque is usually weaker than that of the removed gyro.) If the Vehicle Reference Coordinate Frame (V) associated with lines scribed in the structure of the satellite is now properly relocated by rotation about the principal yaw axis, relative to the deflected Principal Axis Frame (A) described previously, the V frame will again point nominally along the local vertical, the normal to the orbital plane, and the horizontal in the orbital plane. In other words, the Vehicle Reference Frame (V) will then instrument or indicate the Orbital Reference Frame (O) just as it and the Principal Axis Frame (A) did when a full vee configuration of two gyrostabilizers was used.

By the above considerations we are now led to a three-axis damping system that is comprised of only one gyrostabilizer working against the vehicle, which itself exhibits gyroscopic properties. The vehicle also provides the vital gravitational-gradient torque that yields the customary bistable equilibrium orientation in pitch and, depending on the magnitude of the gyrostabilizer's angular momentum, either a monostable or bistable equilibrium orientation in roll. The yaw equilibrium orientation is also either monostable or bistable, depending upon the gyro's angular momentum, but gravitational torque, of course, is not involved. (Normal design would presumably ensure that only a single stable roll and yaw orientation exists, but no device operating only on the gyroscopic principles discussed here is envisioned as capable of providing a single stable pitch orientation.)

By the same process as described above, the removal of one of the gyros of a roll-vee configuration will result in a roll skew of the vehicle principal axes. A single roll-skewed gyrostabilizer system will result. Since a steady state gravitational torque will exist in this case owing to the roll axis deflection of the vehicle's principal axis frame, it is clear from Eq. (2) that the resultant angular momentum of the system (body orbital angular momentum plus gyro angular momentum) must be deflected from $\mathbf{W}_0$ so that its vector time rate of change as seen from inertial space is equal to the vector roll axis gravitational torque acting on the vehicle. That is, the total angular momentum of the system precesses conically in inertial space at orbital frequency about the normal to the orbital plane. It is therefore a nominally constant vector in the V frame (the V frame is appropriately skewed about roll to again instrument the O frame, as before) and is deflected from $\mathbf{W}_0$ in the vertical plane that is normal to the orbital plane. The level of torque required in the torque generator in this case is the magnitude of the body's combined gravitational and gyroscopic restoring torques in the equilibrium orientation.

The fact that the two single gyrostabilizer systems just described actually provide three-axis damping is clarified by the following arguments: As long as the vehicle exhibits roll, pitch, or yaw librations about its equilibrium orientation, either the body or the gyro gimbal—or both—will respond to such librations by relative precession. Any relative motion of the two elements must take place against the viscous restraint of the gyro's damper, and hence results in irreversible energy transfer through heat dissipation. Thus, transient disturbances, which may be represented as discrete quanta of surplus mechanical energy stored initially in the system, must decay as the stored energy is dissipated in the gyrostabilizer damper. Similarly, sinusoidal forcing functions will generally have reduced disturbing effect owing to the dissipative system characteristics.

The two single gyrostabilizer systems are illustrated by Figs. 20 and 21. Each figure shows the equilibrium orientation vector diagram in the plane normal to the skew axis. Figure 20 shows the relationships among the vector quantitites for the single yaw-skewed gyrostabilizer system,

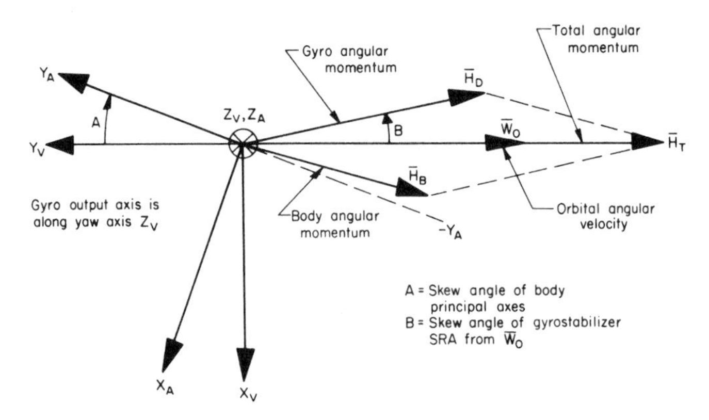

FIG. 20. Vector diagram for equilibrium orientation of single yaw-skewed gyrostabilizer damping system.

and Fig. 21 for the roll-skewed case. Note in Fig. 20 that the total angular momentum $\mathbf{H}_T$ is collinear with the orbital angular velocity $\mathbf{W}_0$ since the yaw deflection of the principal axes produces no gravitational torque. In the roll-skewed case, Fig. 21, the skew of the principal axes about roll gives rise to a roll-axis gravitational restoring torque which requires [by Eq. (2)] that $\mathbf{H}_T$ must have a nonzero time rate of change as

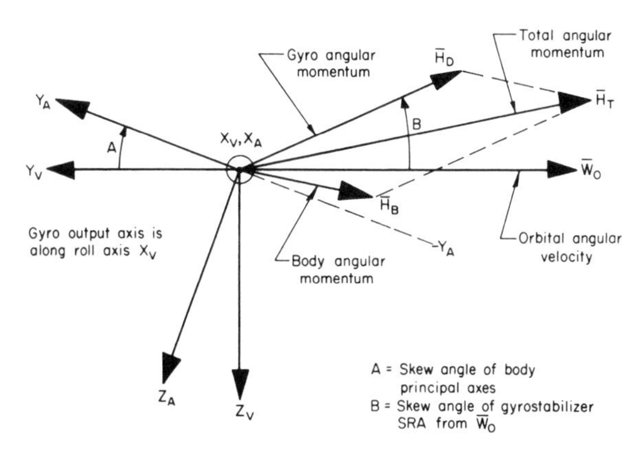

FIG. 21. Vector diagram for equilibrium orientation of single roll-skewed gyrostabilizer damping system.

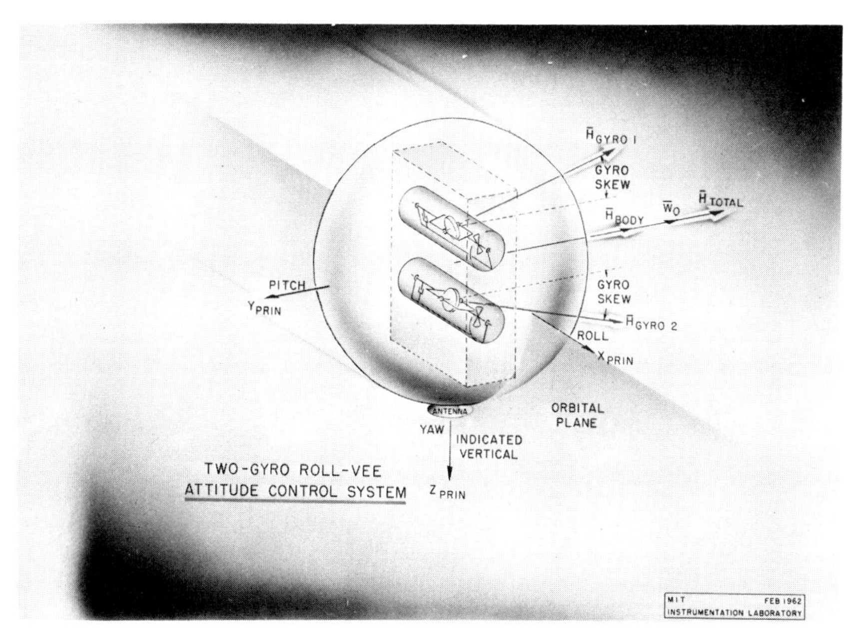

FIG. 22. Two-gyro, roll-vee, attitude-control system.

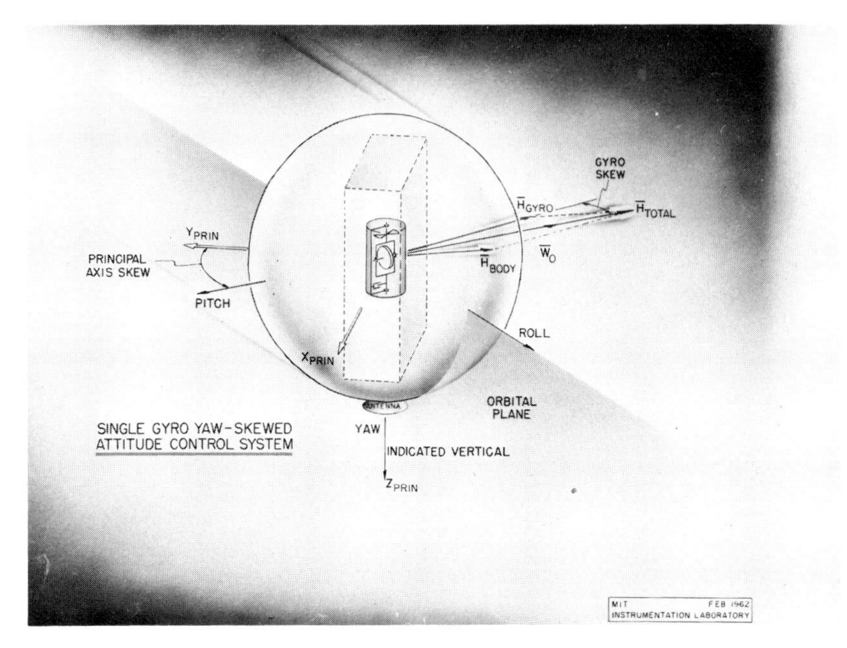

FIG. 23. Single-gyro, yaw-skewed, attitude-control system.

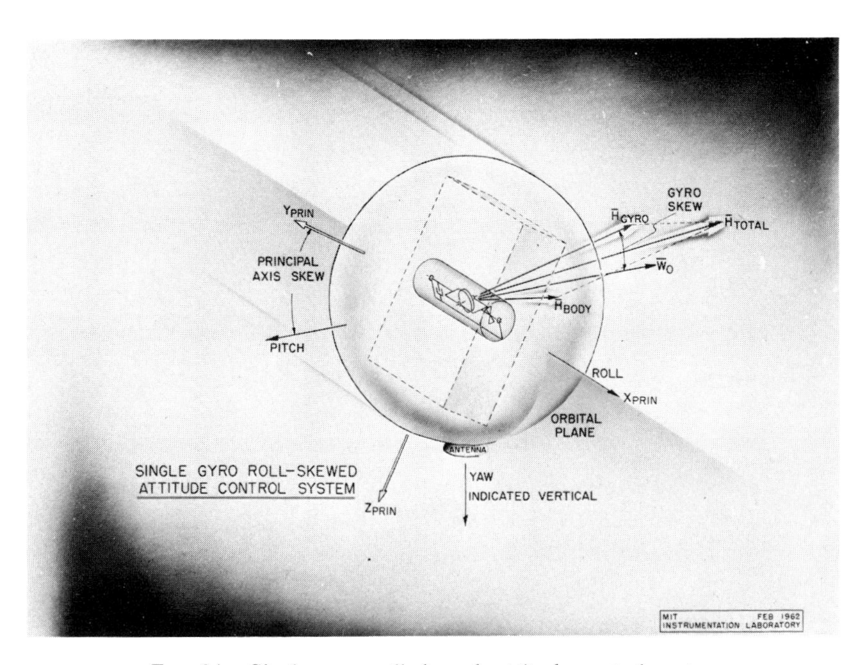

FIG. 24. Single-gyro, roll-skewed, attitude-control system.

seen in inertial space; that is, it cannot be collinear with $\mathbf{W}_0$ but must precess around it in a conical manner at orbital frequency as previously described. The gyrostabilizer skew angle B in each figure is a readily calculable function of the principal axis skew angle A, as is the magnitude of the required torque generator torque to maintain the equilibrium orientation.

Figures 22-24 are presented as a further aid in visualizing the vee and single-gyrostabilizer systems. They depict a small communications satellite, spherical in shape but having an unequal moment-of-inertia distribution to utilize the gravitational-gradient torque effect. In Fig. 22, the satellite is shown with its principal axes unskewed relative to the orbital axes and equipped with two gyrostabilizers arranged in a roll-vee configuration corresponding to Fig. 19. Figure 23 shows the vehicle with principal axes skewed in yaw and equipped with a single yaw-skewed gyrostabilizer as in Fig. 20. The single roll-skewed gyrostabilizer is depicted in Fig. 24, where the vehicle is shown with principal axes skewed about the roll axis as in Fig. 21.

VI. Limitations on Attainable Settling Time

The operation of a gyrostabilizer system in a gravitationally oriented satellite results in the establishment of certain limits to the gyrostabilizer performance. One of these limits, to be discussed here, is the minimum time that such a controlled satellite requires to reach an equilibrium position from given initial conditions. Closely connected to the minimum time problem is the question of whether a gravitationally oriented type of gyrostabilizer satellite can be de-spun from some initial, relatively high, spin rate (e.g., one several times W_0) solely by means of the gyrostabilizer and the gradient of the gravitational field.

Some insight into these questions may be had from energy and momentum considerations. Previous mention was made of the dissipation of the energy of the satellite's libratory motion in the gyrostabilizer damping system. Establishment of limits to settling time based on this energy dissipation may be relatively difficult, however, since there is no absolute assurance that some of the libratory energy is not transferred to another mode (e.g., the orbital mode).

The satellite's angular momentum, on the other hand, seems to offer a simpler and more reliable means of determining the limits to settling time. From the principle of conservation of angular momentum, the total angular momentum of the satellite-earth system remains constant. For the purposes of this investigation, the earth may be treated as a point mass without loss of generality (except for oblateness effects, which are not treated here). Under this assumption, the satellite can exert no torques on the earth. It follows therefore that all of the angular momentum initially possessed by the satellite, including libratory momentum, remains with it.

The momentum of the satellite is exhibited in two forms, i.e., rotation and libration about the vehicle center of mass, and orbital angular momentum. The orbital angular momentum (i.e., the revolution of the satellite mass about the earth) is by far the larger component. When the vehicle is damped, the libratory angular momentum is transferred into orbital angular momentum by means of the gyrostabilizer acting in conjunction with the gravitational field gradient. As long as the gyro gimbal is free to move (not forced to the stops by high rates, for example), the momentum transfer can be effected even for the spinning vehicle having no axis of inertial symmetry. This is because the spin rate for such a vehicle will not be constant, but will be modulated by gravitational-gradient torques.

The maximum torque that can despin or coerce the satellite to the vertical is the maximum torque produced by the gravitational gradient. From Newton's second law, then, the minimum limit to settling time (Δt) that is required to change the angular momentum of spin or libration by an amount ΔH is

$$\Delta t_{(\min)} = \frac{\Delta H}{M_{g(\max)}} \tag{4}$$

In practice and for the gyrostabilizers considered here, this minimum time is not reached. Nevertheless, the bound is useful in determining whether the control technique effected by the gravitational-gradient torque has application to a particular problem.

The considerations of momentum transfer discussed here are generally applicable to those types of vertically oriented satellites that use internal damping and do not depend on mass expulsion, magnetic damping, or other similar means. (For example, magnetic damping offers the possibility of transferring the angular momentum of satellite spin to earth spin momentum.)

VII. Optimization Criteria and Techniques for the Vertical-Vee Configuration

A. Performance Criteria

The preceding sections have demonstrated that a gyrostabilizer system for damping a particular satellite vehicle can be synthesized intuitively.

Once a configuration has been selected, it is necessary to determine the choice of parameters of the system that will maximize its over-all effectiveness for the intended mission.

In order to optimize the system parameters for a particular mission, the criteria to be used as a basis for optimization must be established. Some immediate choices are:

1. Fast settling time (rapid recovery from initial conditions or transients).
2. Minimum pointing error due to sinusoidal disturbances (at orbital or nonorbital frequencies).
3. Minimum peak pointing error due to impulsive torques.
4. Minimum pointing error due to steady torques.
5. Minimum response to orbital eccentricity.

Certain of these requirements may not be entirely consistent with others. For example, the third criterion implies an inertially stiff system that will resist external torque effects, whereas the first implies relatively rapid response to such small external torques as the gravitational-gradient torque. Accordingly, care must be exercised in establishing the initial design specifications in order to minimize the waste of effort that is involved in determination of basic conflicts in the design requirements.

B. Optimization Technique

One approach, and the one recommended here, is to study the effects of certain gyrostabilizer parameters such as angular momentum, gyro damping, and skew angle, on each of the above criteria individually. When each aspect of the problem is thoroughly understood, it becomes possible to view the subject as a whole and to decide on the compromises and trade-offs that might be required. Response to eccentricity, sinusoidal forcing functions, impulses, etc., need no further definitions. A performance index for settling time, however, is needed, there being many definitions for this concept (such as time to $1/e$ of final value, time to 5%, 20% of final value, etc.). More about this subject will follow.

It is fortunate that the motion of a gyrostabilized satellite can be approximated very closely, at least for the balanced vertical-vee configuration under consideration, by a completely linearized set of equations. Thus, in studying the system behavior for angles of less than 10 or 15 deg and for relatively small error rates, the linear equations (sines of angles replaced by angles-usual approximations) hold very closely, particularly for an optimization study.

It is possible to analyze these systems on an analog computer by studying the time solutions to the equations for different parameters. However, a reduction of the problems of transient damping, sinusoidal response, eccentricity response, and steady state response to explicit, closed form, algebraic expressions is warranted for the following reasons:

1. Algebra is a discipline with which all technical personnel are familiar.
2. An algebraic equation is the simplest type to program on a digital computer.
3. The low cost of digital computer evaluation of such equations permits the study of a great number of systems, i.e., design parameter sets.
4. Certain general trends are often discernible from algebraic results.
5. Once the computer programs are written, the optimization of any control system describable by the same set of equations can be completed in a few days.

With regard to the need for a specific settling time criterion, a transient performance index has been suggested by Mr. Mark A. Smith of the Instrumentation Laboratory, Massachusetts Institute of Technology. It is the integral of the square of the error response to analytic transients (impulses, steps, etc.) that are applied to the vehicle and damping system. (The use of this index is treated at length in reference [22].) Since the error is squared before the integration process, the minimization of this particular index tends to penalize systems having large peak errors and systems which permit relatively small errors to persist for an appreciable time. Experience has shown that a fairly good correlation exists between minimization of actual settling time by analog or other techniques and minimization of the integral squared error. The principal advantage of the integral squared error index is the resulting efficiency of the optimization process. Closed form algebraic solutions are obtained by analytic manipulation. These equations need not be factored for their roots. They can be readily programmed for digital computation with sufficient flexibility to accommodate wide parameter ranges of the vehicle, the orbit, the damping system, and the disturbances. Once the set of algebraic equations is obtained and programmed, a brief computational effort using a modern, high-speed digital computer will yield enough data points (i.e., evaluated integral squared errors for particular system configurations and disturbance inputs) to plot a complete set of curves from which optimum configurations may be readily determined.

Thus, the first and third (the transient) criteria noted above may be applied using the integral squared error techniques just described. The periodic and steady disturbances can also be studied using similar analytic methods in conjunction with the linearized equations, and combined with efficient digital evaluation of the closed form solutions. Algebraic solutions are again obtained for the system responses to eccentricity and to sinusoidal and steady external torques. These are evaluated digitally for a wide range of parameters. Further families of optimization curves are thus obtained showing the optimum systems for these disturbances. Once equipped with a full set of these curves for a particular mission, the designer is able to examine trade-offs in system performance and arrive rather quickly at a system configuration offering the optimum compromises among the various performance indices. He is assured by the scope of his information, as presented in the compact form of his design curves, that his choice of an optimum system is well founded. Actual simulation of the selected system, using analog or digital techniques, will generally be carried out to verify system performance and to determine actual transient responses. If the design effort is carefully applied, this simulation should support the choice of system parameters.

An interesting further simplification of this design approach is suggested by the closed form of the solutions obtained for the various disturbances. It is particularly applicable to a relatively long, slender satellite having an inertia distribution approximating that of a dumbbell-shaped body. The solutions in their general form may be differentiated with respect to an appropriate parameter. It is then possible, using elementary methods for determining minima and maxima of a function, to determine that value of the selected parameter yielding the minimum point on the associated design curve. By this means one can, in a short time and with no machine computation required, determine the approximate vertical-vee gyrostabilizer system parameters needed for the particular vehicle and mission. The system parameters so selected will be reasonably near to those selected as a result of detailed optimization. Hence, this technique is useful in making preliminary design estimates with fair accuracy.

A useful consequence of the linear nature of the system equations for small errors and error rates is as follows: Optimization of the damping system parameters for a given vehicle at a given altitude may be extended simply and directly to the same vehicle at any other altitude, provided the same normalized performance criteria and disturbance environment are imposed. This extension is accomplished by simply changing the angular momentum H by the same percentage and in the same direction as the change in W_0. The values of gyro gain (H/C) and gyro case skew angle (γ) will not change, nor will the settling time if measured in orbital periods. It should be noted that certain factors can render this simple extension inapplicable. Portions of the disturbance environment (e.g., aerodynamic torque, solar torque, micrometeoritic bombardment) may be radically different at the various altitudes considered. This might cause a shift in emphasis in system design from, say, minimization of deflections due to steady torques (at lowest altitudes) to minimization of response to periodic forcing functions. Any change in performance requirements will, of course, necessitate a re-optimization.

A detailed exposition of the analytic optimization techniques discussed here is to be found in reference [8].

C. Illustration of the Analytic Optimization Techniques

To illustrate the use of analytic techniques in designing a vertical-vee gyrostabilizer system for a particular mission, examples of certain of the curves that result will be presented. Assume $I_X = 10{,}000$ slug-ft², $I_Y = 10{,}000$ slug-ft², $I_Z = 3333$ slug-ft², and $W_0 = 6.25 \times 10^{-4}$ radians/sec.

The first curve, Fig. 25, shows I_θ, a measure of pitch integral squared

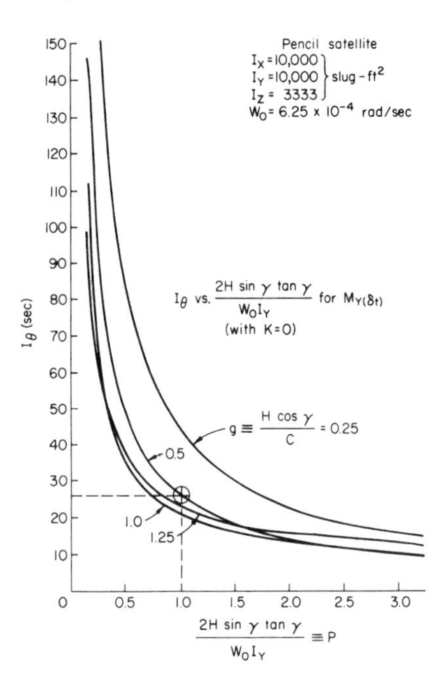

FIG. 25. Normalized integral square error in pitch (I_θ) for 1 ft-lb-sec torque impulse in pitch ($M_{Y(\delta t)}$).

error for an impulse torque about the pitch axis, plotted against P, a normalized angular momentum that is effective in damping pitch. The parameter g is a measure of the effective damping ratio of each gyroscope.

The effective angular momentum of the gyrostabilizer system is normalized to the vehicle angular momentum in pitch since it has been found, as observed in previous sections, that the optimum effective gyro angular momentum for good roll and yaw response is of the order of $W_0 I_Y$, the satellite orbital angular momentum.

Figure 26 shows a similar set of curves for an impulsive torque in roll. The abscissa in this case is R, a measure of the normalized effective gyro angular momentum which damps the roll-yaw system.

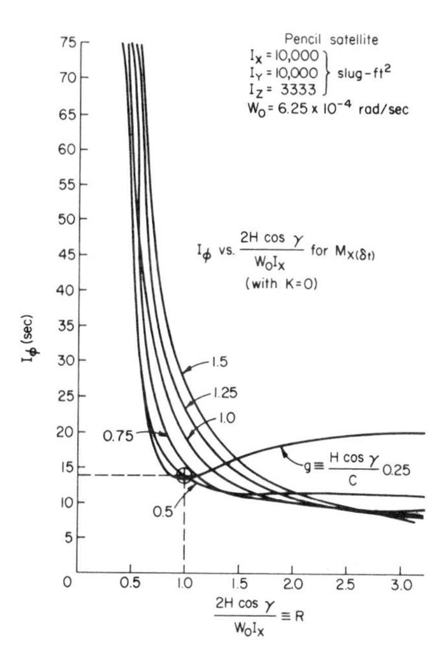

FIG. 26. Normalized integral-square error in roll (I_φ) for 1 ft-lb-sec torque impulse in roll ($M_{X(\delta t)}$).

The curves for step torques in pitch and roll are shown as Figs. 27 and 28, respectively. They are seen to be similar to the previous curves for impulses. Note that any value for P or R of unity or greater provides rather effective damping. The system response to a sinusoidal forcing function at W_0 (Fig. 29) and to orbital eccentricity (Fig. 30) will now be examined. (Response at W_0 may be of importance since for many orbits solar torques have a large component at this frequency. $2W_0$ may also be

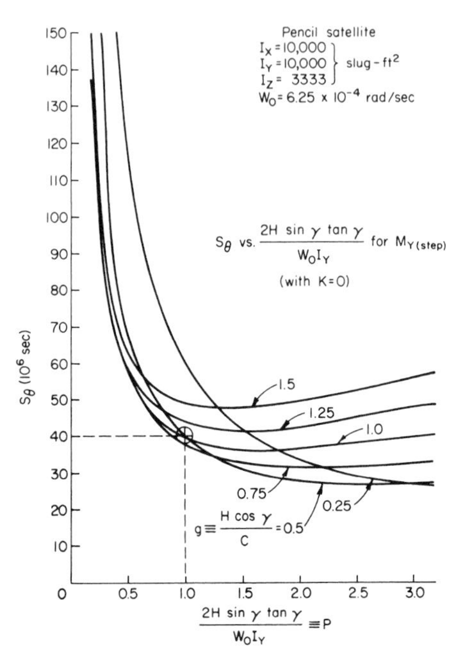

Fig. 27. Normalized integral-square error in pitch (S_θ) for 1 ft-lb step torque in pitch ($M_{Y(\mathrm{step})}$).

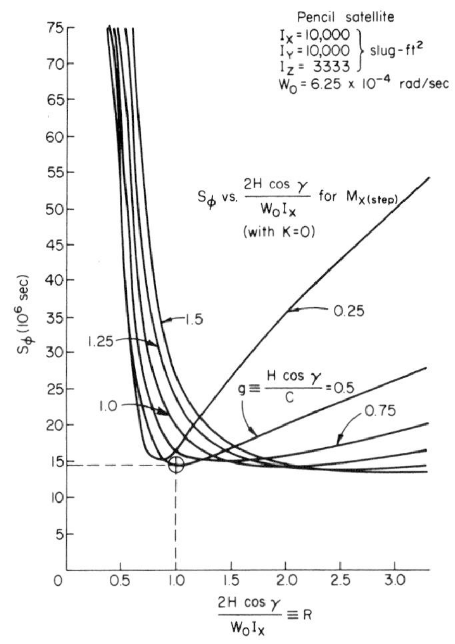

FIG. 28. Normalized integral-square error in roll (S_φ) for 1 ft-lb step torque in roll ($M_{X(\mathrm{step})}$).

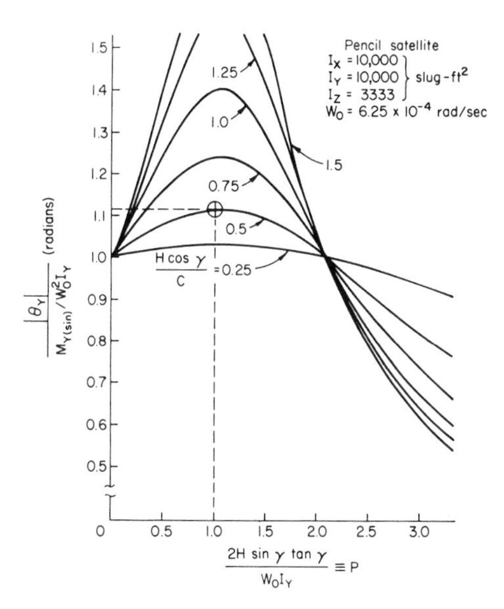

Fig. 29. Normalized pitch response to an external pitch axis sinusoidal forcing function ($M_{Y(\sin)}$) at orbital frequency.

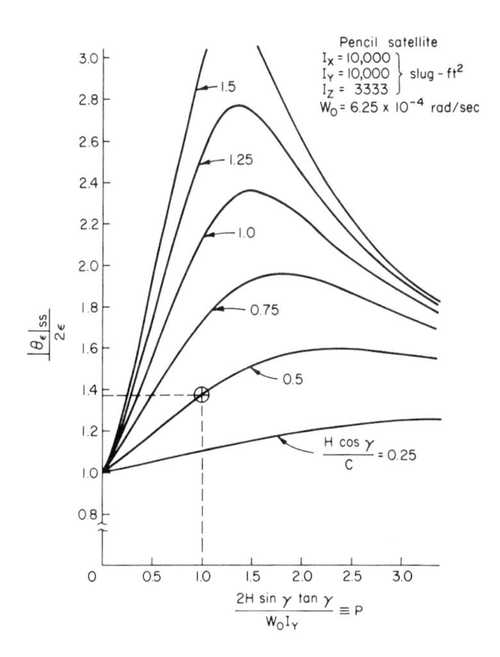

Fig. 30. Normalized pitch response to eccentricity.

important for the same reason.) Both of these curves show that large values of $(H/C) \cos \gamma$ are to be avoided for values of P in the range of 1 to 2 or 3. A compromise system tending to favor both good damping and low response to sinusoids at orbital frequency and eccentricity is marked on the curves at $P = R = 1$ and $g = 0.5$. If we use the defining values for P, R, and g, the result is the following gyrostabilizer system parameters. For each gyro:

$$H = 4.4 \text{ slug-ft}^2 \text{ sec}^{-1}$$

$$= 4.4 \text{ lb-ft-sec}$$

$$H/C = 0.7$$

$$|\gamma| = 45 \text{ deg}$$

With such a system selected, a more detailed examination may then be made, including time solutions to impulses and initial conditions and frequency response in all axes to periodic forcing functions. Analog and/or digital computers will be useful for such studies.

Note, however, that without such further study a considerable amount of information can be obtained from the optimization curves. From the integral squared error curves it is apparent that very nearly the best transient damper has been selected for the particular satellite. Past studies have shown that for such a "best" damper choice, initial transients will be reduced to 1/5 their maximum value in one to two orbits.

Figure 30 shows $|\theta_e|/2\epsilon = 1.37$. Thus, for $\epsilon = 0.01$ as an example, the maximum deflection of the long axis of the vehicle from the vertical about the pitch axis due to eccentricity will be 0.0274 radians or about 1.6 deg.

Figure 29 shows maximum pitch error response of approximately 0.025 radian (1.45 deg) per 10^{-4} ft-lb of peak amplitude of a sinusoidal pitch torque at orbital frequency.

Various modifications of the method described are suggested for selecting parameters of a vertical-vee gyrostabilizer system. For example, if a particular pair of gyros is available, having some value of angular momentum within range of the optimum value, adjustment of the skew angle γ and possibly of the damping ratio H/C may permit achievement of the desired characteristics. Certain modifications in the inertia distribution of the satellite may also be possible. The referenced study [8] contains the derivation of equations required to plot the necessary curves for the vertical-vee configuration.

VIII. Further Satellite Applications for Gyrostabilizers

In the attitude control systems discussed in the preceding sections, gyrostabilizers have been employed to achieve solutions to the stabilization problem in a particular class of satellite vehicles, namely, the earth-oriented satellites stabilized relative to the rotating orbital reference frame *O*. The comments presented are based upon actual optimization experience related to this class of vehicles. Relatively less effort has been directed by the present authors to the use of gyrostabilizers in other types of missions. However, some of the work of others can be noted here, and some further applications possibly deserving of study can be suggested.

Gyrostabilizers as defined and described in this paper are not commanded; i.e., their precessional motions are inherent consequences of the rotational environment in which they are placed. If a set of such instruments is instead torqued by command signals and constrained by high gain feedback to respond only to such signals, then, in effect, the set of "gyro-like" devices can be the equivalent of a set of inertia reaction flywheels. Specifically, any pair of these "gyro-like" devices oriented back-to-back as described previously (Fig. 16), but operated in the fully commanded mode, will be functionally equivalent to an inertia reaction wheel with rotational axis along the axis normal to their spin and precession axes, provided that the momentum saturation capabilities are equivalent in the two cases. The advantage of two constant-speed gimbaled wheels over a single ungimbaled reaction wheel in such a case is moot, at best, since external sensing and computation would be required in either method. On the other hand, if integrating gyroscopes are permitted to precess without command as gyrostabilizers and, in addition, are fed open-loop torquing commands from external sensors in the manner of reaction wheels, it is possible to devise systems having apparent advantages over their purely reaction-wheel counterparts. Such arrangements have been studied [9, 10, 19, 20] and are being studied [21].

An area of possible investigation in this latter regard is to utilize the outputs of roll and pitch horizon scanners to torque the gyrostabilizers of a yaw-vee configuration in order to obtain additional roll and pitch stiffening of the vehicle. The pitch scanner signal would be applied equally, but oppositely, to the gyro torquers to move their angular momentum vectors in a "scissor" mode, thus producing a pitch axis torque. The roll scanner signal would be applied to precess the angular momentum vectors equally in a constant vee angle or "antiscissor" mode, which would result—for the yaw-vee configuration—in a net torque about the roll axis. Owing to the integrating nature of the gyroscopes, the pitch and roll torques so produced would be proportional to the sensed pitch and roll error angles, respectively, and, within the saturation capabilities of the instruments, would augment the inherent gravitational restoring torques of the vehicle. The normal semipassive action of the gyrostabilizers should achieve the necessary damping. As in all of the gyrostabilizer applications, gyroscopes with relatively large gimbal rotational freedom (e.g., ± 10 to ± 30 deg) should be used with systems of the latter type.

In connection with the scissor torquing described above, it is worth noting that application of such torques to the gyrostabilizers of the vertical-vee configuration in the form of a periodic, constant, low-amplitude "dither" signal of relatively high frequency (e.g., about ten to a hundred times orbital frequency) will overcome threshold problems of the gyros. Stiction and coulomb friction can be completely overcome, and gyros of the unfloated type, having drift rates even as high as earth rate, can be made to perform as well as perfect gyros as far as the capability of sensing vehicle rotations is concerned. The power requirement connected with dithering is one to two orders of magnitude below that required in applying the constant gimbal torque in maintaining the basic vee angle of the control system. Note that, in systems employing torquing commands to the gyrostabilizers from sensors such as horizon scanners, it is probable that the random noise components often present in the output of such devices will generally eliminate the need for any additional dithering.

The classical method of stabilizing a space vehicle, and one that was widely used in early spacecraft launchings, is spin stabilization. One of the serious problems associated with such vehicles is the damping of nutations (i.e., wobbling motions) of the axis of spin. Another is the need for maintaining a fixed spatial orientation of the spin axis over extended periods. An interesting further application for gyrostabilizers is in the solution of the nutation problem. A single gyrostabilizer would be mounted in the spinning vehicle, with its spin reference axis parallel to the vehicle's spin axis (presumably the axis of maximum moment of inertia, which is the only unconditionally stable spin axis); this gyrostabilizer would dissipate nutational energy through its gimbal motion relative to its case. The equilibrium condition for such a system should occur only when the spin axes of the vehicle and gyrostabilizer are coincident and nutation of the vehicle's spin axis has ceased.

The problem of erecting a vehicle spin axis to a preferred spatial orientation and/or of maintaining that spatial orientation cannot be solved by gyrostabilizer techniques alone. This is evident from Eq. (2) which requires that any change in the net angular momentum of the closed system, comprised of the spinning vehicle and its internal gyrostabilizer, must be produced by the application of an unbalanced torque external to the closed system. Such a torque may be derived in a variety of ways that have received considerable attention in the literature [17]. If one assumes that an erection torque is available, then damping of transients due to erection and impulsive disturbances, as well as the basic damping of nutation, can be accomplished by the single gyrostabilizer oriented as previously noted.

An interesting possibility exists for erecting the spin axis of an equatorial meteorological or similar satellite to parallelism with the orbital angular momentum vector (i.e., the normal to the orbital plane). This is achieved by configuring the vehicle with an inertia distribution similar to that of a disk having a thickness that is significantly less than its radius. Such a vehicle will experience a gravitational-gradient restoring torque about the horizontal axis in the orbital plane whenever the vehicle spin axis deviates, about that horizontal axis, from the normal to the plane. Unfortunately, there are at least two basic limitations on this torque. First, it is very small; yet it is required to precess the net angular momentum vector of the vehicle and gyrostabilizer, which is rather large. Long time constants will almost certainly be involved at best. (See Section VI.) Second, for a fixed orientation of the spin axis in space that is nonparallel with the normal to the orbital plane, the small roll axis gravitational-gradient torque will vary sinusoidally about a zero mean in opposition to the roll error itself, which will also be sinusoidal with zero mean. (The misalignment interchanges sinusoidally between the orbital roll and yaw axes as the vehicle circles the earth with no inertial rotation other than its spin). There is, of course, no yaw axis gravitational-gradient torque, so erection about the yaw axis would have to occur as a consequence of gyrocompassing in the manner previously described. One means of accomplishing the desired erection is to inject the satellite into its orbit either spinning at orbital rate or at least at a very low rate. The erection would be accomplished in a relatively short time (e.g., of the order of a few orbital periods) by the gravitational-gradient torque and the damping action of the gyrostabilizer. Once erection is complete and transients have decayed sufficiently, the vehicle could be spun up about the indicated normal to the orbital plane. Any subsequent transients or nutations induced during spin-up would be damped by the gyrostabilizer.

The above remarks concerning further applications of gyrostabilizers are intended to suggest areas needing further study. The present authors are unaware of any flight testing related to these methods. It is recommended, however, that consideration be given to the use of gyrostabilizers as sensing, computing, and control elements in future spacecraft mission design studies to determine their merits relative to alternative means.

IX. Summary and Conclusions

An attempt has been made to promote a more general awareness of the application of gyrostabilizers to a vertically oriented satellite vehicle, and to show how control systems using these devices may be synthesized by intuitive processes. The mathematical analyses supporting these systems are not included here, but have been carried out, both by the present authors and by others as referenced. Optimization studies have demonstrated the feasibility of the use of semipassive integrating gyros as attitude control devices. Such systems can easily meet orientational requirements of the order of a few degrees. Because of the inherent simplicity of the systems, they are well suited to applications in which longevity is a prime factor. In view of the relatively simple analytic optimization techniques that are presently available for the vertical-vee configuration, and are derivable for other arrangements, analytic complexity in applying gyrostabilizers has been significantly reduced.

Although hardware aspects of the gyrostabilizer systems are not emphasized here, it is pertinent to extract some comments in this regard from the cited references by the present authors. Specifically, the bearing problems of gyroscopes used in the acceleration environment of orbital motion have not as yet been fully defined, but it seems reasonable that, with the possible exception of self-acting, gas-lubricated spin bearings (owing to whirl resonance phenomena), the extreme reduction of g loadings in bearings in the satellite may significantly increase their useful life. The power requirements of the gyro systems are expected to be relatively low as compared with active systems offering similar damping. System responses to transient and sinusoidal disturbances are affected somewhat by variations in gyro H/C ratio, (as shown by the curves presented in this paper), but it appears likely that temperature control may be a minor problem or no problem at all, even for the floated types of integrating gyros. When specification tolerances are relatively severe, gyros are available having compensation means to maintain the H/C ratio constant.

Of the several systems described, the vee configurations appear to be most attractive owing to their superior damping capabilities. The single-gyro roll-skewed system may be competitive with a vee system in certain applications owing to its simplicity, but it still would have inherent limitations compared with the vee configurations, particularly in regard to response to orbital eccentricity. For any proposed application, it is as necessary with these systems as with any other to weigh all factors in order to ensure selection of the proper system.

References

1. R. E. Roberson, Attitude control of a satellite vehicle—an outline of the problems, *7th Intern. Astronaut. Congr. Barcelona, 1957.*
2. R. E. Roberson, Gyroscopic control of satellite attitude. *Proc. 1st Symp. Intern. Rockets and Astronautics, Tokyo, Japan, 1959.*
3. M. G. Kaye, Attitude Control of Satellites Using Integrating Gyros. M. S. Thesis T-208, Instrument. Lab. Mass. Inst. Technol., 1959.
4. E. G. Burt, On the attitude control of earth satellites. *8th Anglo-Am. Aeronaut. Conf. London, September, 1961.* Roy. Aeronaut. Soc., London, England.
5. G. Ogletree, S. J. Sklar, and J. G. Mangan. Satellite Attitude Control Study, *Instrument. Lab. Mass. Inst. Technol., Rept. R-308, Part I,* July 1961.
6. Satellite Attitude Control Study, *Instrument. Lab. Mass. Inst. Technol. Rept. R-308, Part II,* February 1962.
7. J. De Lisle, G. Ogletree, and B. M. Hildebrant, Attitude Control of Satellites Using Integrating Gyroscopes, *Instrument. Lab. Mass. Inst. Technol. Rept. R-350,* December 1961.
8. B. M. Hildebrant, J. Lombardo, and T. Petranic, Analytic Techniques Applied to Satellite Gyrostabilizers, *Instrument. Lab. Mass. Inst. Technol. Rept. R-398,* February 1963.
9. R. C. Wells, J. S. Sicko, and F. M. Courtney, Gyroscopic Low Power Attitude Control for Space Vehicles, *Tech. Doc. Rept. No. ASD-TDR-62-580.* Flight Control Lab., Aeronaut. Systems Div., Air Force Systems Command, Wright-Patterson Air Force Base, Ohio, September 1962.
10. J. S. White and Q. M. Hansen, Study of a Satellite Attitude Control System Using Integrating Gyros as Torque Sources, *NASA Tech. Note, NASA TN D-1073.* National Aeronautics and Space Administration, Washington, D. C., September 1961.
11. Frederic I. Ordway III, ed. *Advances in Space Sci.* **2**, (1960).
12. W. Wrigley, R. B. Woodbury, and J. Hovorka, Inertial Guidance, *Sherman M. Fairchild Publ. Fund Paper No. FF-16.* Inst. Aeronaut. Sci., New York, 1957.
13. R. E. Roberson, Gravitational torques on a satellite vehicle. *J. Franklin Inst.* **265**, No. 1, p. 13, January 1958.
14. L. Page, "Introduction to Theoretical Physics," 3d ed. Van Nostrand, Princeton, New Jersey, 1952.
15. H. Goldstein, "Classical Mechanics." Addison-Wesley, Reading, Massachusetts, 1959.
16. C. S. Draper, W. Wrigley, and L. Grohe, The Floating Integrating Gyro and Its Application to Geometrical Stabilization Problems on Moving Bases, *Sherman M. Fairchild Publ. Fund Paper No. FF-13.* Inst. of Aeronaut. Sci., New York, 1955.
17. R. E. Roberson, Methods for the Control of Satellites and Space Vehicles, Vol. I, Sensing and Actuating Methods, and Vol. II, Control Mechanization and Analysis, *Wright Air Develop. Div. Tech. Rept. 60-643,* July 1960.
18. R. E. Roberson, Torques on a satellite vehicle from internal moving parts. *J. Appl. Mech. 25,* 196-200 (1958).
19. R. H. Cannon, Jr., Gyroscopic Coupling in Space Vehicle Attitude Control Systems, *ASME Paper No. 61-JAC-8,* 1961.

20. A. G. Buckingham, A new method of attitude control utilizing the earth's magnetic field for long life space vehicles. *Nat. Conf. on Guidance, Control and Navigation, Am. Rocket Soc. Stanford Univ., Palo Alto, Calif., August, 1961.*
21. N. C. New, Spacecraft Attitude Control for Extended Missions. ScD Thesis, Mass. Inst. Technol., Cambridge, Massachusetts, to be published.
22. G. Newton, L. Gould, and J. Kaiser, "Analytical Design of Linear Feedback Controls," Wiley, New York, 1957.

Generalized Gravity-Gradient Torques

ROBERT E. ROBERSON

Department of Engineering, University of California, Los Angeles, California

I. INTRODUCTION

WITHIN THE PAST FEW YEARS there has been a spate of literature on the problem of gravity-gradient torques on satellite vehicles. Actually, as Russell A. Nidey has pointed out to me, the essential results were anticipated by Tisserand in his *Traité de mécanique céleste* in 1891. The recent works seem to involve three basic approaches: the use of potential methods (e.g., reference [1]); vector methods (e.g., references [2-5][1]); derivation of torque expressions taking account of the oblateness of the attracting body (e.g., references [1, 2]). All of these seem to be limited to first-order results in the ratio of satellite dimension to its distance from the center of attraction, a very small number for normal vehicles. Except for the third category, all are confined to simple inverse square gravitational fields.

As regions of space are reached where lunar and solar gravitational effects become more significant, it becomes important to relax the last restriction and to consider more general cases. In particular, it is interesting to examine some of the simple general properties of torque in an inverse square field, to see how sensitive they are to changes in the field structure and, if they carry over in some degree to more general fields, how the behavior is modified. How much can one say in general

[1] Only when preparing the present note did this author discover that the results of reference [5] are just those of Eq. 76 in reference [3].

about the role of the normal to the equipotential surface and the equilibrium orientation of the body?

II. Basic Relationships

Nidey's basic result (Eq. (7) of reference [4]) can be rewritten in the form

$$\bar{L} = C\bar{e}_3 \times \mathbf{I}^* \cdot \bar{e}_3 \tag{1}$$

where $\bar{L}$ is the torque on the body, $\bar{e}_3$ is the unit vector in the outward radial direction from the single gravitating center, $\mathbf{I}^*$ is related to the inertia dyadic $\mathbf{I}$ (as shown in reference [5]) by

$$\mathbf{I}^* = \tfrac{1}{2}(\operatorname{diag} \mathbf{I})\,\mathbf{E} - \mathbf{I} \tag{2}$$

$\mathbf{E}$ is the unit dyadic, and C is a function of the instantaneous position of the center of mass of the vehicle (a constant with respect to attitude motions). Consider two basic properties of this representation as guides to what to seek in the general case.

Theorem I

No torque component exists along $\bar{e}_3$.

Proof: Form the scalar product $\bar{L} \cdot \bar{e}_3$. It vanishes since $\bar{e}_3 \cdot \bar{e}_3 \times$ (any vector) $= 0$.

Theorem II

If $\bar{e}_1$, $\bar{e}_2$ unit vectors are adjoined to $\bar{e}_3$ to form a right-handed orthogonal triad, the former defining the plane normal to $\bar{e}_3$ through the vehicle's center of mass, a sufficient condition that $\bar{L} = 0$ is that $\bar{e}_1$, $\bar{e}_2$, $\bar{e}_3$ be a set of principal axes for $\mathbf{I}^*$ and hence for $\mathbf{I}$. A necessary condition that $\bar{L} = 0$ is that $I_{13} = I_{23} = 0$ referred to $\bar{e}$, $\bar{e}_2$, $\bar{e}_3$.

Proof: If $\bar{e}_1$, $\bar{e}_2$, $\bar{e}_3$ are body principal axes, $\mathbf{I}^*$ is diagonal. Then $\bar{e}_3 \times \mathbf{I}^* \cdot \bar{e}_3 = \bar{e}_3 \times I^*_{i3}\bar{e}_i = \bar{e}_3 \times I^*_{33}\bar{e}_3 = 0$, proving sufficiency. If $\bar{L} = 0$, $I^*_{i3}\bar{e}_3 \times \bar{e}_i = (I^*_{13}\bar{e}_2 - I^*_{23}\bar{e}_1) = 0$ whence $I^*_{13} = I^*_{23} = 0$, equivalent to $I_{13} = I_{23} = 0$, proving the necessary condition.

It might have been anticipated in Theorem II that there would be no requirement that $I_{12} = 0$ since Theorem I already assures that neutral stability exists about $\bar{e}_3$. Another viewpoint is that the orientation of $\bar{e}_1$, $\bar{e}_2$ about $\bar{e}_3$ can be chosen arbitrarily without affecting the equilibrium state, as should be the case because of the rotational symmetry of the field itself about $\bar{e}_3$.

Now consider the more general case. It is assumed only that the gravitational potential function φ is locally smooth enough to permit an expansion about the origin of $\bar{e}_1$, $\bar{e}_2$, $\bar{e}_3$ in the form

$$\varphi = \varphi_0 + a_1 x + a_2 y + a_3 z + \tfrac{1}{2} b_{11} x^2 + b_{12}\, xy + \tfrac{1}{2} b_{22}\, y^2 + b_{13}\, xy + b_{23}\, yz + \tfrac{1}{2} b_{33}\, z^2 + \cdots \qquad (3)$$

where x, y, z are coordinates along $\bar{e}_1$, $\bar{e}_2$, $\bar{e}_3$, respectively, measured from the origin. The requirement that $\bar{e}_3$ be normal to the equipotential surface is equivalent to

$$\left.\frac{\partial\varphi}{\partial x}\right|_0 = \left.\frac{\partial\varphi}{\partial y}\right|_0 = 0$$

or $a_1 = a_2 = 0$. If the local field strength is f_0, $a_3 = -f_0$. The vector force $\bar{f}^{(\alpha)}$ on the αth element of the body is the negative gradient of the potential function, or

$$f^{(\alpha)} = f_0 \bar{e}_3 - \mathbf{B} \cdot \bar{\rho}^{(\alpha)} \qquad (4)$$

where $\mathbf{B}$ is the symmetric dyadic $b_{ij}\bar{e}_i\bar{e}_j$ and $\bar{\rho}^{(\alpha)} = x\bar{e}_1 + y\bar{e}_2 + z\bar{e}_3$, with x, y, z referring to the location of the αth element.

The total torque on the body, if m_0 is the mass of the αth element, is

$$\bar{L} = \sum_\alpha m_\alpha \bar{\rho}^{(\alpha)} \times \bar{f}^{(\alpha)} = \sum_\alpha m_\alpha\, \bar{\rho}^{(\alpha)} \times [f_0 \bar{e}_3 - \bar{\rho}^{(\alpha)} \times \mathbf{B} \cdot \bar{\rho}^{(\alpha)}] \qquad (5)$$

the first sum on the right-hand side of Eq. (5) vanishes if the coordinate is taken as the center of mass of the body and one is left with

$$\begin{aligned} \bar{L} &= -\sum_\alpha m_\alpha \bar{\rho}^{(\alpha)} \times \mathbf{B} \cdot \bar{\rho}^{(\alpha)} \\ &= \sum_\alpha m_\alpha b_{ij} \bar{e}_i \times \bar{\rho}^{(\alpha)} \bar{\rho}^{(\alpha)} \cdot \bar{e}_j \\ &= b_{ij} \bar{e}_i \times \mathbf{I}^* \cdot \bar{e}_j \end{aligned} \qquad (6)$$

Equation (6) could have been obtained simply by a power expansion of the vector function $\bar{f}^{(\alpha)}$ about the origin, but it is convenient to have the interpretation of the b_{ij} as the coefficients of the quadratic terms in the local potential function.

III. Some General Properties

Perhaps the first question that arises in connection with this general representation concerns the conditions under which it reduces to the same form as for the inverse square field case. It can be settled by the following theorem.

Theorem III

A necessary and sufficient condition that the general representation Eq. (6) reduce to the inverse square field representation Eq. (1) is that **B** be diagonal with $b_{11} = b_{22}$.

Proof: The condition is sufficient. If **B** is diagonal, Eq. (5) gives

$$\begin{aligned}\bar{L} &= b_{11}(I_{21}^{*}\bar{e}_3 - I_{31}^{*}\bar{e}_2) + b_{22}(I_{32}^{*}\bar{e}_1 - I_{12}^{*}\bar{e}_3) + b_{33}(I_{13}^{*}\bar{e}_2 - I_{23}^{*}\bar{e}_1)\\ &= (b_{22} - b_{33})\, I_{23}^{*}\bar{e}_1 + (b_{33} - b_{11})\, I_{13}^{*}\bar{e}_2 + (b_{11} - b_{22})\, I_{12}^{*}\bar{e}_3.\end{aligned}$$

If $b_{11} = b_{22}$ this reduces to

$$\bar{L} = (b_{11} - b_{33})\,(I_{23}^{*}\bar{e}_1 - I_{13}^{*}\bar{e}_2) = (b_{11} - b_{33})\,\bar{e}_3 \times \mathbf{I}^{*} \cdot \bar{e}_3.$$

The condition is also necessary. If there is to exist a structure of **B** and value of C such that $b_{ij}\bar{e}_i \times \mathbf{I}^{*} \cdot \bar{e}_j = C\bar{e}_3 \times \mathbf{I}^{*} \cdot \bar{e}_3$, these parameters must satisfy the three scalar equations derived from the latter, namely,

$$b_{ij}I_{2j}^{*} - b_{2j}I_{ij}^{*} = 0 \tag{7a}$$

$$-b_{ij}I_{3j}^{*} + b_{3j}I_{ij}^{*} = CI_{13} \tag{7b}$$

$$b_{2j}I_{3j}^{*} - b_{3j}I_{2j}^{*} = -CI_{23} \tag{7c}$$

Furthermore, since these must be identities regardless of the values of I_{ij}^{*}, they must hold for diagonal $\mathbf{I}^{*}$ with unequal elements. With the latter choice, Eq. (7a) implies $b_{12} = b_{21} = 0$, Eq. (7b) implies $b_{13} = b_{31} = 0$, Eq. (7c) implies $b_{23} = b_{32} = 0$; i.e., **B** must be diagonal, and this must then be true for any other choice of $\mathbf{I}^{*}$. If we let **B** be diagonal and use the symmetry of $\mathbf{I}^{*}$, Eq. (7b) implies $b_{33} = b_{11} = C$ while Eq. (7c) implies $b_{33} - b_{22} = C$, i.e., $b_{11} = b_{22}$, and the condition of necessity is established.

From Theorem III it is seen that the fundamental difference of the general case is having off-diagonal terms in **B** or $b_{11} \neq b_{22}$. It now remains to be seen what can be said under these conditions. An important question is whether one can ever achieve the conditions of Theorem III (i.e., **B** diagonal with $b_{11} = b_{22}$) for a system of more than one point mass. The following theorem is proved for an arbitrary number of point masses, but can be generalized in a somewhat different form to distributed systems.

Theorem IV

For a set of more than one gravitating center, a necessary and sufficient condition that $b_{11} = b_{22}$ is that all the masses be collinear with the center of mass of the body. Collinearity is also a sufficient condition for the diagonality of **B**.

Proof: Only a necessity is proved, the proof of sufficiency being obvious. Consider a collection of particles of which a representative one is shown in Fig. 1. We are interested in the form of the potential function

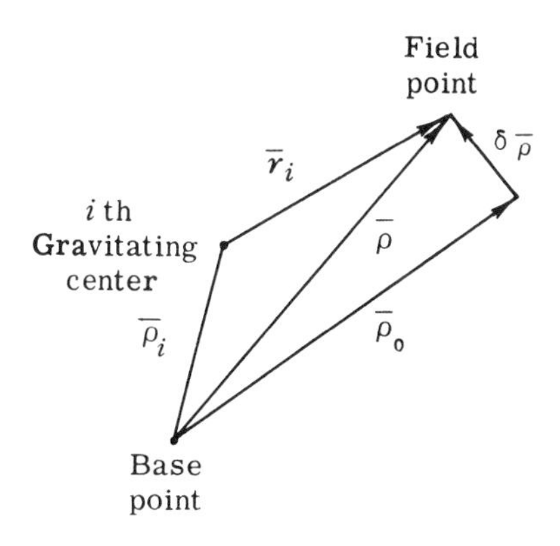

FIG. 1.

at the field point (characterized by $\bar{\rho}$) in the neighborhood of a specified $\bar{\rho}_0$ which, as before, is to be identified with the center of mass of the body. Letting $\bar{\rho} = \bar{\rho}_0 + \delta\bar{\rho}$ and retaining only second-order terms in the components of $\delta\bar{\rho}$ [consistent with Eq. (3)], the potential can be written

$$\begin{aligned}\varphi &= -\sum_i \frac{K_i}{r_i} = -\sum_i \frac{K_i}{|\bar{\rho}_0 - \bar{\rho}_i + \delta\bar{\rho}|} \\ &= -\sum_i \frac{K_i}{|\bar{\rho}_0 - \bar{\rho}_i|} + \sum_i \frac{K_i(\bar{\rho}_0 - \bar{\rho}_i)\cdot\delta\bar{\rho}}{|\bar{\rho}_0 - \bar{\rho}_i|^3} \\ &\quad + \tfrac{1}{2}\sum_i \frac{K_i\{|\bar{\rho}_0 - \bar{\rho}_i|^2\,|\delta\bar{\rho}|^2 - 3\,[(\bar{\rho}_0 - \bar{\rho}_i)\cdot\delta\rho]^2\}}{|\bar{\rho}_0 - \bar{\rho}_i|^5} + \cdots\end{aligned}$$

Let $\bar{e}_1$, $\bar{e}_2$, $\bar{e}_3$ be an arbitrarily oriented set of orthogonal base vectors centered at $\bar{\rho}_0$. Write $\delta\bar{\rho} = x\bar{e}_1 + y\bar{e}_2 + z\bar{e}_3$, $\bar{\rho}_0 - \bar{\rho}_i = \bar{R}_i = R_{ix}\bar{e}_1 + R_{iy}\bar{e}_2 + R_{iz}\bar{e}_3$ with $R_i = |\bar{R}_i|$. Then the quadratic terms in x, y, z are derived simply. The condition $b_{11} = b_{22}$ becomes

$$\sum_i \frac{K_i(R_{iy}^2 + R_{iz}^2 - 2R_{ix}^2)}{R_i^5} = \sum_i \frac{K_i(R_{ix}^2 + R_{iz}^2 - 2R_{iy}^2)}{R_i^5} = 0$$

But this is equivalent to

$$\sum_i \frac{K_i(R_{iy}^2 + R_{ix}^2)}{R_i^5} = 0$$

Since this is a sum of squares equal to zero, it must vanish term by term; i.e., $R_{iy} = R_{ix} = 0$. This implies that the $\bar{e}_3$ axis of the local frame must be simultaneously parallel to all of the vectors $\bar{R}_i$. This can happen only when all masses and the field point are collinear, and the condition of necessity is proved.

IV. Equilibrium States

We come now to the basic problem of equilibrium conditions in a general field. The general representation of Eq. (6) is used and the scalar product $\bar{L} \cdot \bar{e}_q$ is formed ($q = 1, 2, 3$). One has

$$\bar{L} \cdot \bar{e}_q = b_{ij} I_{mn}^* (\bar{e}_i \times \bar{e}_m)\, (\bar{e}_n \cdot \bar{e}_j) \cdot \bar{e}_q = \epsilon_{imq} b_{ij} I_{mj}^*$$

The condition $\bar{L} \cdot \bar{e}_q = 0$ is then equivalent (in both necessary and sufficient sense) to

$$b_{2j} I_{3j}^* - b_{3j} I_{2j}^* = 0 \qquad (q = 1) \tag{8a}$$

$$b_{3j} I_{1j}^* - b_{1j} I_{3j}^* = 0 \qquad (q = 2) \tag{8b}$$

$$b_{1j} I_{2j}^* - b_{2j} I_{1j}^* = 0 \qquad (q = 3) \tag{8c}$$

These results provide direct proof of the following theorems.

Theorem V

If the body is so oriented with respect to $\bar{e}_1$, $\bar{e}_2$, $\bar{e}_3$ that **I** (hence **I***) is diagonalized, then a necessary and sufficient condition that there be no torque about the e_3 axis (the gravity-gradient direction) is that either $b_{12} = 0$ or $I_{11}^* = I_{22}^*$.

Proof: Using the diagonality of **I**, Eq. (8c) becomes $b_{12}(I_{11}^* - I_{22}^*) = 0$ and the theorem follows.

The interpretation of this result is simple. If the body is aligned with one principal axis along the gravity-gradient direction, then the gravitational torque about this direction vanishes if and only if either the body is symmetric about this axis, or the two orthogonal axes of symmetry of the field itself coincide with the two remaining principal axes of the vehicle.

Similar results, of course, hold for torques about the other two axes, but this first case is of special importance because it is the analog of Theorem I. It should not be necessary to prove here that there is at least one orientation of $\bar{e}_1$, $\bar{e}_2$ which makes $b_{12} = 0$, this being a basic property of quadratic forms.

Theorem VI

If the three principal moments of inertia are unequal, the only equilibrium configurations are those for which the body principal axes coincide with the principal axes of the dyadic **B**.

Proof: It follows from Theorem V that the torque vanishes about $\bar{e}_3$ if and only if $b_{12} = 0$ since $I_{11}^* \neq I_{22}^*$. Similar theorems hold for axes $\bar{e}_1$ and $\bar{e}_2$, whence a necessary and sufficient condition for equilibrium is that $b_{12} = b_{23} = b_{31} = 0$. But this is precisely the condition that $\bar{e}_1$, $\bar{e}_2$, $\bar{e}_3$ be principal axes of **B**.

Note that the equilibrium state is not unique. For example, a physical body with unequal principal moments will be in equilibrium if $I_{33} > I_{22} > I_{11}$ or $I_{22} > I_{11} > I_{33}$ or in any other combination. But the stability of equilibrium will depend on which of these inequalities hold, and that is the next problem to be considered.

V. Stability of Equilibrium

Consider that $\bar{e}_1$, $\bar{e}_2$, $\bar{e}_3$ are principal axes for the dyadic **B**. The torque components L_i along these axes become [analogously with Eqs. (8)]

$$L_1 = (b_{22} - b_{33})\, I_{23}^* \tag{9a}$$

$$L_2 = (b_{33} - b_{11})\, I_{13}^* \tag{9b}$$

$$L_3 = (b_{11} - b_{22})\, I_{12}^* \tag{9c}$$

Let α_{ij} be direction cosines which relate principal body axes $\bar{X}_1$, $\bar{X}_2$, $\bar{X}_3$ to $\bar{e}_1$, $\bar{e}_2$, $\bar{e}_3$ according to $\bar{X}_i = \alpha_{ij}\bar{e}_j$. Since the inertia dyadic in principal axes is

$$\mathbf{I} = I_1\bar{X}_1\bar{X}_1 + I_2\bar{X}_2\bar{X}_2 + I_3\bar{X}_3\bar{X}_3 \tag{10a}$$

it follows that in the $\bar{e}_j$ frame,

$$\mathbf{I} = (I_1\alpha_{1j}\alpha_{1k} + I_2\alpha_{2j}\alpha_{2k} + I_3\alpha_{3j}\alpha_{3k})\, \bar{e}_j\bar{e}_k \tag{10b}$$

According to Eq. (2), the off-diagonal terms of $\mathbf{I}^*$ are precisely the negatives of the off-diagonal terms of $\mathbf{I}$ itself.

Now we need certain properties of direction cosine matrices. If we confine attention to very small deviations of the $\bar{X}_i$ from the $\bar{e}_i$ frame for purposes of stability analysis, the diagonal elements α_{ii} are unity and the off-diagonal terms are of first order in small quantities, to within second-order errors in these quantities. It is well known that one may focus on the linear approximation for an analysis of stability in the neighborhood of neighborhood of equilibrium, the nonlinear (i.e., second-order) terms playing a role only if the linear approximation vanishes identically. Finally, because of the orthogonality of $\bar{X}_i$ and $\bar{X}_j$ $(i \neq j)$, one has relations of the form $\sum_k \alpha_{ik}\alpha_{jk} = 0\ (i \neq j)$.

In view of the above remarks, one has to first order in small quatities

$$I_{23}^* = -(I_2\alpha_{22}\alpha_{23} + I_3\alpha_{32}\alpha_{33}) + \cdots$$

and $\alpha_{22}\alpha_{23} + \alpha_{32}\alpha_{33} +$ second-order terms $= 0$, whence

$$I_{23}^* = (I_2 - I_3)\,\alpha_{32} + \cdots \tag{11a}$$

Similarly,

$$I_{13}^* = (I_1 - I_3)\,\alpha_{31} + \cdots \tag{11b}$$

$$I_{12}^* = (I_1 - I_2)\,\alpha_{21} + \cdots \tag{11c}$$

Then the torque components in Eqs. (9) approach the following values for small deviations from equilibrium:

$$L_1 = (b_{22} - b_{33})\,(I_2 - I_3)\,\alpha_{32} + \cdots \tag{12a}$$

$$L_2 = (b_{33} - b_{11})\,(I_1 - I_3)\,\alpha_{31} + \cdots \tag{12b}$$

$$L_3 = (b_{11} - b_{22})\,(I_1 - I_2)\,\alpha_{21} + \cdots \tag{12c}$$

Now $\alpha_{32} > 0$ means that $\bar{X}_3$ projects positively on $\bar{e}_2$ and that $L_1 > 0$ implies a restoring torque under the displacement. Similarly, $\alpha_{31} > 0$ means that the $\bar{X}_3$ axis projects positively on $\bar{e}_1$ and that $\bar{L}_2 < 0$ is required to restore the body to equilibrium. Finally, $\alpha_{21} > 0$ requires $L_3 > 0$ for stability about the $\bar{e}_3$ axis.

All of these results can be collected as a theorem:

Theorem VII

Necessary and sufficient conditions for stability in the neighborhood of equilibrium when $\bar{e}_1$, $\bar{e}_2$, $\bar{e}_3$ are principal axis for $\mathbf{B}$ are:

$$(b_{22} - b_{33})\,(I_2 - I_3) > 0 \text{ for stability about } \bar{e}_1$$

$$(b_{11} - b_{33})\,(I_1 - I_3) > 0 \text{ for stability about } \bar{e}_2$$

$$(b_{11} - b_{22})\,(I_1 - I_2) > 0 \text{ for stability about } \bar{e}_3$$

For an inverse square field about one gravitating center it is easy to show that $b_{33} = -3b_{11} = -3b_{22} < 0$. For any other field similar in the sense that both $b_{22} - b_{33} > 0$ and $b_{11} - b_{33} > 0$, it follows that $\bar{e}_3$ is a position of stable equilibrium if and only if both $I_3 < I_2$ and $I_3 < I_1$. That is, $\bar{X}_3$ must be the body axis with the smallest moment of inertia. This is precisely the common result for an inverse square field. One can experiment conceptually with arrangements of masses which vary the elements b_{ij}, but it does not seem easy to give a general answer to the question of the conditions (on mass distribution) under which both $b_{22} - b_{33} > 0$ and $b_{11} - b_{33} > 0$. In the remaining case of stability about $\bar{e}_3$, if $b_{11} > b_{22}$ it must be that $I_1 > I_2$ for stability. This can be interpreted easily by reference to the Fig. 2 which shows the intersection

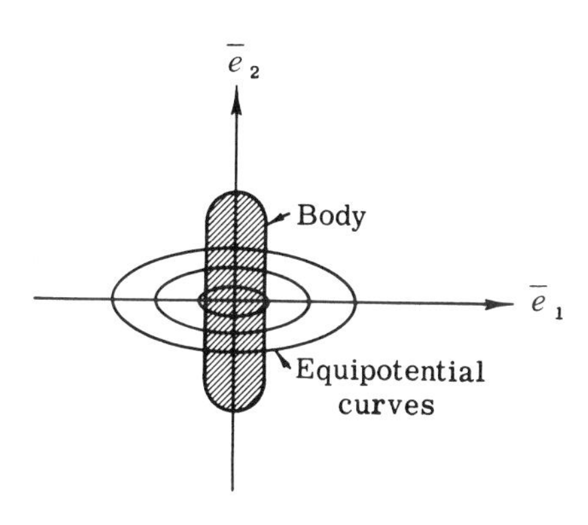

FIG. 2.

of several of the equipotential surfaces with the $\bar{e}_1\bar{e}_2$ plane, drawn for the condition $b_{11} > b_{22}$ (both positive). Then the body cross section for $I_1 > I_2$ has the disposition shown in the shaded ellipse. Again, the "long axis" of the body (cross-sectioned in the $\bar{e}_1\bar{e}_2$ plane) tends to line up with the direction of maximum field gradient in that same plane.

VI. Torques

The remaining question of some interest is the form of the torque components themselves. Two cases are especially useful: components relative to $\bar{e}_1$, $\bar{e}_2$, $\bar{e}_3$ and those relative to $\bar{X}_1$, $\bar{X}_2$, $\bar{X}_3$. The general torque expression, of course, is given by Eq. (6). The components relative to $\bar{e}_1$, $\bar{e}_2$, $\bar{e}_3$ may be deduced from Eqs. (9), or in the small-angle case take the form of Eq. (12). It is simple to prove from Eq. (6) that if **B** is diagonal with respect to $\bar{e}_1$, $\bar{e}_2$, $\bar{e}_3$, then the body axis torque components are

$$\bar{L} \cdot \bar{X}_1 = (I_2 - I_3)(b_{ii}\alpha_{2i}\alpha_{3i}) \tag{13a}$$

$$\bar{L} \cdot \bar{X}_2 = (I_3 - I_1)(b_{ii}\alpha_{3i}\alpha_{1i}) \tag{13b}$$

$$\bar{L} \cdot \bar{X}_3 = (I_1 - I_2)(b_{ii}\alpha_{1i}\alpha_{2i}) \tag{13c}$$

A third set sometimes is of interest, namely a mixed set consisting of $\bar{X}_3$, an axis in the $\bar{e}_1\bar{e}_2$ plane normal to $\bar{X}_3$, and a third axis to make a right-hand orthogonal frame. The torque components in this set are not given here explicitly, but they are straightforward enough to develop and it can be shown that, unlike the simple inverse square field case, the torque component along the third axis does not vanish in general.

References

1. R. E. Roberson, Gravitational torque on a satellite vehicle. *J. Franklin Inst.* **265**, 13-22 (1958).
2. D. DeBra, Vectors and dyadics: The natural tools of space-vehicle mechanics. *Am. Astronaut. Soc., San Francisco, August 1961.*
3. R. M. Frick and T. B. Garber, General Equations of a Satellite in a Gravitational Field, *Rand Rept. RM-2527,* December 1959.
4. R. A. Nidey, Gravitational torque on a satellite of arbitrary shape. *ARS J. 30,* 203-4 (1960).
5. R. E. Roberson, Alternate form of a result by Nidey. *ARS J.* **11**, 1292-1293 (1961).

Aerodynamic and Radiation Disturbance Torques on Satellites Having Complex Geometry

WILLIAM J. EVANS

Grumman Aircraft Engineering Corporation,
Bethpage, New York

I. INTRODUCTION

TWO OF THE SIGNIFICANT CONTRIBUTORS to the attitude disturbance of a satellite are the aerodynamic flow and radiation flux fields. At altitudes on the order of 100 miles or more, the atmosphere is highly rarefied to the point where collisions between gas molecules in a unit of volume become negligibly few. However, there are sufficiently many molecules present in a unit of volume to determine macroscopic gas properties. The large mean distance that a molecule travels between collisions, its mean free path, permits the assumption that there is no interaction between the incident stream of molecules and the reflected or re-emitted stream. The incident free stream is thus effectively undisturbed by the presence of a body, and the aerodynamic forces may be calculated by considering separately the incident and reflected molecular flows.

At higher altitudes, where the aerodynamic forces tend to become exceedingly small, another disturbance phenomenon emerges. The impact of solar radiation assumes equal importance. The incident electromagnetic energy of radiation imposes a pressure and a shear stress upon an intercepting surface. This follows from both the electromagnetic and quantum theories of light.

One might suppose that a satellite can be symmetrically designed, and that aerodynamic and radiation disturbance torques may thus be avoided. However, many satellites require solar paddles and other geometric complications, which, when shielding is taken into account, lead to severe asymmetry. Shielding is a major factor in the production of disturbance torques. It is therefore deserving of considerable attention in any attempt to predict these torques.

II. Aerodynamic Analysis

In free molecule flow the interaction of the molecular flux with a surface is formulated on the basis of coefficients. The coefficients relate the extent to which the properties of the incident flow are accommodated to the conditions of the surface, as well as the nature of the re-emission flow pattern. These coefficients are the surface reflection coefficients for tangential and normal momentum (σ, σ'), and the thermal accommodation coefficient (α):

$$\sigma = \frac{\tau_i - \tau_r}{\tau_i}, \qquad \sigma' = \frac{P_i - P_r}{P_i - P_w}, \qquad \text{and} \qquad \alpha = \frac{E_i - E_r}{E_i - E_w} \tag{1}$$ [1]

When these coefficients are zero there is no energy or tangential momentum exchange between the incident stream and the surface, and the reflection is said to be specular (Fig. 1). When these coefficients have

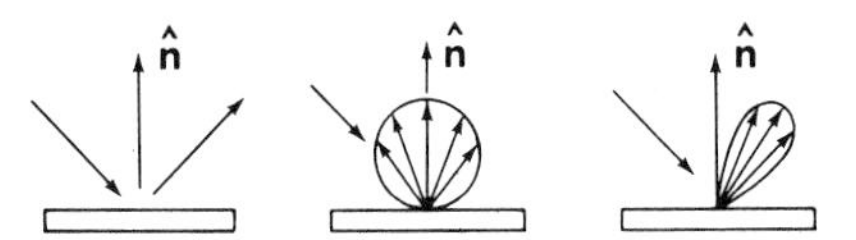

Fig. 1. Reflection geometry. (a) Specular; (b) diffuse; (c) Partially diffuse.

the value of unity ($\alpha = \sigma = \sigma' = 1.0$) the molecules are re-emitted randomly, and in complete thermodynamic equilibrium with the surface; whereupon the reflection process is said to be completely diffuse. The coefficients may therefore be thought to represent the percentage of the reflected molecules that are re-emitted diffusely.

In the past, available data has appeared to indicate values of σ and α of approximately 0.9 to 1.0. Although no workable method had been devised for measuring σ', it was supposed that its values lay at the same end of the spectrum. More recently, molecular beam experimentation has indicated

[1] For definitions of symbols used in this chapter, see Appendix.

a downward shift in these values, and this with apparatus not yet capable of simulating the orbital environment, and which should be expected to yield values higher than are to be expected under orbital conditions. With improved laboratory technique enabling better simulation of orbital conditions it is entirely possible that all prior data will have been discredited. The coefficients may be found to vary greatly from one another, as functions of surface temperature, speed, and incidence. Therefore in appraising the accuracy attainable in any delicate engineering application, one must consider the validity of the basic data, based upon present-day low-density testing capability, and the extreme difficulty involved in attempting to simulate orbital conditions. It is also questionable, in applying the surface accommodation coefficients, to assume them to be constant, independent of speed, incidence, and surface temperature. Other representations have been proposed to replace these accommodation coefficients, but little corroborating data have been offered to support any other system, or to show that it represents an advancement in terms of the assumptions required and the resulting phenomenological validity. One can thus perceive the need for further development along these lines. Experimental and theoretical research on molecule-surface interaction is being carried out under NASA contract (NASr-104) at Grumman Aircraft Engineering Corporation. The satellite aerodynamic environment is simulated by using a shock tube to introduce a molecular beam of high energy into a test chamber maintained at an extremely low pressure, and methods of measuring distributions of reflected-molecule flux and energy are under development.

A local coordinate system is defined on an element of surface so that x is the inward normal, and the relative velocity vector $\mathbf{U}$ is in the x, y plane

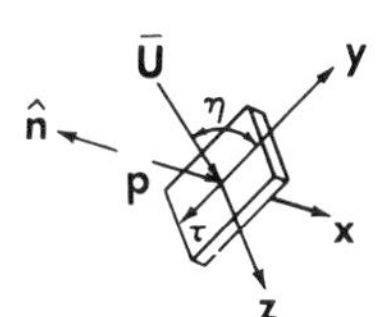

FIG. 2. Local coordinate system.

(Fig. 2). The incident free stream flow is assumed to be in Maxwellian equilibrium, so the molecular distribution function is given by:

$$f = \frac{\rho}{m} (2\pi RT)^{-3/2} \exp \left\{ \frac{-1}{2RT} [(u - u_0)^2 + (v - v_0)^2 + (w)^2] \right\} \tag{2}$$

The distribution function f has the property that $f d\tau \, d\omega$ represents the number of molecules in a volume element $d\tau$ having velocities within an element $d\omega$ of the velocity space, or thus the number of molecules in an element $d\tau\, d\omega$ of the phase space. The normal momentum flux results in a pressure which is expressed as the sum of the pressures due to the incident and re-emitted flux of momenta.

$$P = P_i + P_r \tag{3}$$

The uncertain re-emitted flux must be represented in terms of the surface reflection coefficient σ'; then:

$$P = (2 - \sigma')\, P_i + \sigma' P_w \tag{4}$$

Integrating over all possible values of the velocity space, we find that

$$P_i = \int_{-\infty}^{\infty} mu^2 f\, d\omega, \qquad P_w = \int_{-\infty}^{\infty} mu^2 f_w\, d\omega$$

$$P_i = \frac{\rho U^2}{2\sqrt{\pi}\, s}\{(s \sin \eta) \exp[-(s \sin \eta)^2] + \sqrt{\pi}\,[\tfrac{1}{2} + (s \sin \eta)^2]\,[1 + \operatorname{erf}(s \sin \eta)]\} \tag{5}$$

and

$$P_w = \tfrac{1}{2}\rho R \sqrt{TT_w}\{\exp[-(s \sin \eta)^2] + \sqrt{\pi}(s \sin \eta)\,[1 + \operatorname{erf}(s \sin \eta)]\}$$

where erf (x) is the probability integral,

$$\operatorname{erf}(x) = \frac{2}{\sqrt{\pi}} \int_0^x \exp(-x^2)\, dx \qquad 0 \leqslant \operatorname{erf}(x) < 1 \tag{6}$$

The complete pressure relation is thus,

$$P = \frac{\rho U^2}{2s^2}\left\{\begin{aligned} &\exp[-(s')^2]\left[\frac{2-\sigma'}{\sqrt{\pi}}(s') + \frac{\sigma'}{2}\sqrt{\frac{T_w}{T}}\right] \\ &+ [1 + \operatorname{erf}(s')]\left\{(2-\sigma')\,[\tfrac{1}{2} + (s')^2] + \frac{\sigma'}{2}\sqrt{\frac{\pi T_w}{T}}\,(s')\right\} \end{aligned}\right. \tag{7}$$

where $s' = s \sin \eta$. It is seen that the pressure is a complicated function of surface temperature, T_w, and the reflection coefficient, σ'.

The shear stress acting upon a surface is the amount of tangential momentum transmitted to the surface by the flow. It is, therefore, expressed as the difference between the incident and re-emitted values of the tangential momenta

$$\tau = \tau_i - \tau_r \tag{8}$$

Again, the re-emitted momentum flux must be obtained through the reflection coefficient relation, and thus,

$$\tau = \sigma \tau_i \tag{9}$$

Integrating over the velocity space, we have

$$\tau_i = - \int_{-\infty}^{\infty} (muv) f \, d\omega$$

and (10)

$$\tau = \frac{\sigma \rho U^2}{2\sqrt{\pi}\, s} (\cos \eta) \{\exp [-(s \sin \eta)^2] + \sqrt{\pi} (s \sin \eta) [1 + \operatorname{erf} (s \sin \eta)]\}$$

The shear stress is directly proportional to the reflection coefficient, σ, and completely independent of the wall temperature, T_w. This is, of course, due to the fact that completely diffuse re-emission is symmetric and imparts no tangential momentum to resist that imparted to the surface by the incident flux ($\tau_w = 0$).

Variation of pressure and shear stress with incidence for completely diffuse reflection is shown in Fig. 3 for a satellite in a circular orbit at

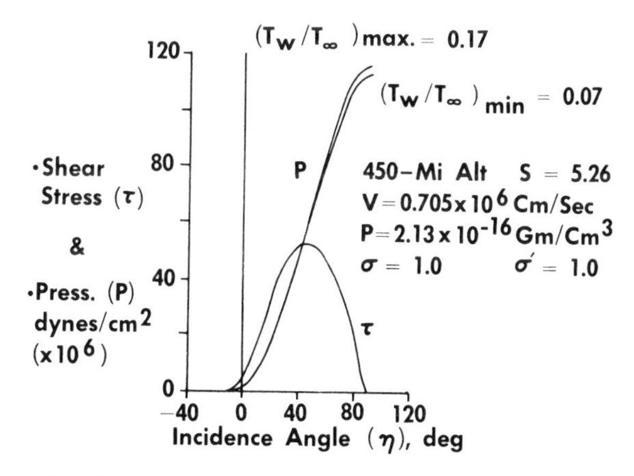

FIG. 3. Aerodynamic pressure and shear stress vs. incidence.

450 miles altitude. The radiation equilibrium surface temperature is shown to be an order of magnitude less than the free stream temperature, and to produce a negligible effect upon the surface pressure due to variation over the working range of temperature. In this condition (that is, $T_w \ll T$), the pressure may easily be approximated to be directly proportional to the factor $(2 - \sigma')$, and so the pressure is magnified linearly as the coefficient is decreased, until finally, for pure specular

reflection, it is doubled. The shear stress reduces linearly to zero as the reflection becomes specular.

Figure 4 illustrates the severe variation of density with altitude and the significant uncertainty of this basic physical parameter. The dashed portions of the curves represent extrapolations and, therefore, regions of doubt. The pressure and shear stress will thus be rather sensitive to altitude variation, and oblateness of the atmospheric density profiles.

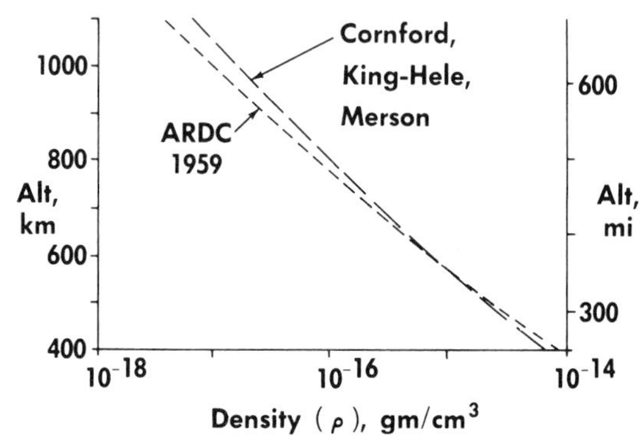

FIG. 4. Density vs. altitude.

The analysis presented thus far assumes convex surface character. That is, the reflected molecular flux is nowhere considered to reimpinge upon another element of surface. For satellites having complex geometry this may appear not to be a very good approximation. However, let us consider the phenomenon. The free stream flow has a high ordered velocity with a superimposed Maxwellian distribution of random molecular velocities. All of this momentum is absorbed by a surface element at initial impact. If the subsequent re-emission is completely diffuse, it will have associated with it a momentum flux P_w, which is orders of magnitude less than the incident flux. If the re-emission is not completely diffuse, its momentum flux components will be given by

$$P_r = (1 - \sigma') P_i + \sigma' P_w$$

and (11)

$$\tau_r = (1 - \sigma) \tau_i$$

and these can obviously have some significant value. However, as long as the re-emission behaves somewhat diffusely in its mechanical velocity profile, the number flux of re-emitted molecules that will intercept another element of surface is inversely proportional to the square of the distance between the elements in question. The effects of convex geometry are thus minimized until the limit of purely specular reflection is approached.

In light of this, and of the previously mentioned uncertainties, the effect of concave geometry is at best clearly of second order.

III. Radiation Analysis

The incident electromagnetic energy of radiation imposes a pressure and a shear stress upon an intercepting surface. Consider a flux of radiant energy crossing unit area in unit time (E in thermal units or S in mechanical units). By Einstein's mass-energy equivalence

$$S = mc^2 \tag{12}$$

where m represents the mass of radiant energy, and c, the speed of light. The momentum flux may now be stated as:

$$mc = \frac{S}{c} \tag{13}$$

For an intercepting plate that is arbitrarily oriented, the incident energy is given by:

$$S_i = S \cos \theta \tag{14}$$

and the incident momentum flux is:

$$(mc)_i = \frac{S_i}{c} = \frac{S}{c} \cos \theta \tag{15}$$

The resulting pressure and shear stress components are

$$P_i = \frac{S}{c} \cos^2 \theta$$

and

$$\tau_i = \frac{S}{c} \cos \theta \sin \theta \tag{16}$$

Continuity demands that, at the intercepting surface,

$$\rho + \tau' + \alpha = 1 \tag{17}$$

where

ρ = reflectivity
τ' = transmissivity
α = absorptivity

For a well-insulated surface at equilibrium

$$\tau' = 0 \qquad \text{and} \qquad \alpha = 1 - \rho \tag{18}$$

For radiation re-emitted diffusely according to the cosine law

$$(P_r)_d = \rho \frac{S}{c} \cos\theta \frac{1}{\pi} \int_{\varphi=0}^{\varphi=2\pi} \int_{\theta'=0}^{\theta'=\pi/2} \cos^2\theta' \sin\theta' \, d\theta' \, d\phi \tag{19}$$

where ϕ is the meridian angle of the sphere of re-emission.

$$(P_r)_d = \frac{2}{3} \rho \frac{S}{c} \cos\theta \tag{20}$$

For purely specular re-emission, however,

$$(P_r)_s = \rho \frac{S}{c} \cos^2\theta. \tag{21}$$

The balance of the radiation is absorbed and invisibly emitted in a diffuse pattern:

$$P_e = \frac{2}{3} (1 - \rho) \frac{S}{c} \cos\theta \tag{22}$$

The resulting pressure therefore is given by two different relations, covering the cases of completely diffuse and purely specular reflection:

$$P = P_i + P_r + P_e$$

Diffuse:

$$P_d = \frac{S}{c} \cos\theta \left[\cos\theta + \frac{2}{3}\right] \tag{23}$$

Specular:

$$P_s = \frac{S}{c} \cos\theta \left[\cos\theta + \frac{2}{3} + \rho \left(\cos\theta - \frac{2}{3}\right)\right]$$

In the case of diffuse reflection, the tangential momentum of re-emission, as well as that of emission, is zero, i.e.,

$$(\tau_r)_d = \tau_e = 0 \tag{24}$$

However, for specular reflection,

$$(\tau_r)_s = \rho \frac{S}{c} \cos\theta \sin\theta \tag{25}$$

The tangential momentum flux transmitted to the surface is the difference between the incident and all re-emitted portions:

$$\tau = \tau_i - (\tau_r + \tau_e)$$

Diffuse:

$$\tau_d = \frac{S}{c} \cos\theta \sin\theta \tag{26}$$

Specular:

$$\tau_s = (1 - \rho) \frac{S}{c} \cos\theta \sin\theta$$

As with the aerodynamic analysis, one is able to formulate coefficients expressing the percentage of radiation that is re-emitted diffusely:

$$A = \frac{P_i - P_r/\rho}{P_i - P_w/\rho} \quad \text{and} \quad B = \frac{\tau_i - \tau_r/\rho}{\tau_i} \tag{27}$$

For completely diffuse reflection:

$$P_r = P_w, \quad \tau_r = \tau_w, \quad A = B = 1$$

For purely specular reflection:

$$\frac{P_r}{\rho} = P_i, \quad \frac{\tau_r}{\rho} = \tau_i, \quad A = B = 0$$

The equations may then be combined in the form:

$$P = \frac{S}{c} \cos\theta \left[\left(\cos\theta + \frac{2}{3}\right) + \rho(1 - A)\left(\cos\theta - \frac{2}{3}\right)\right]$$

and

$$\tau = \frac{S}{c} \cos\theta \sin\theta[1 - \rho(1 - B)] \tag{28}$$

This provides a means for examining the effect of varying the amount of diffuse reflection. However, these coefficients serve only as a formulation artifice here, for no experimental data exist for them. In the aerodynamic analysis the incident and reemitted number flux are one and the same. ($\rho_{\text{aero}} = 1$). In the radiation analysis, the existence of at least ρ and α in the continuity relation lead to the necessity of somehow measuring ρ and one of the coefficients simultaneously.

The radiation sources to be considered for an Earth satellite are the

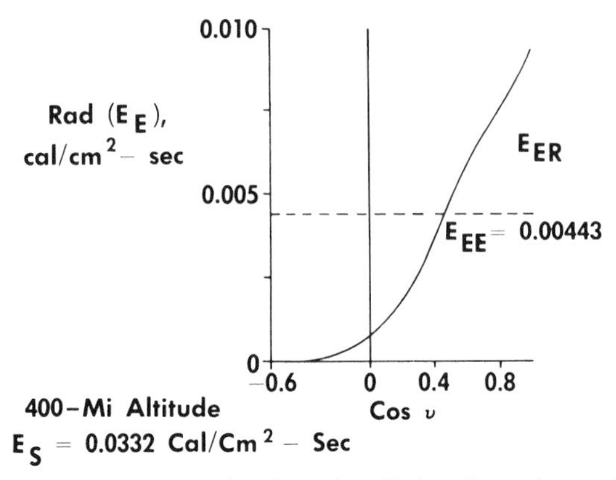

FIG. 5. Emitted and reflected radiation from the earth.

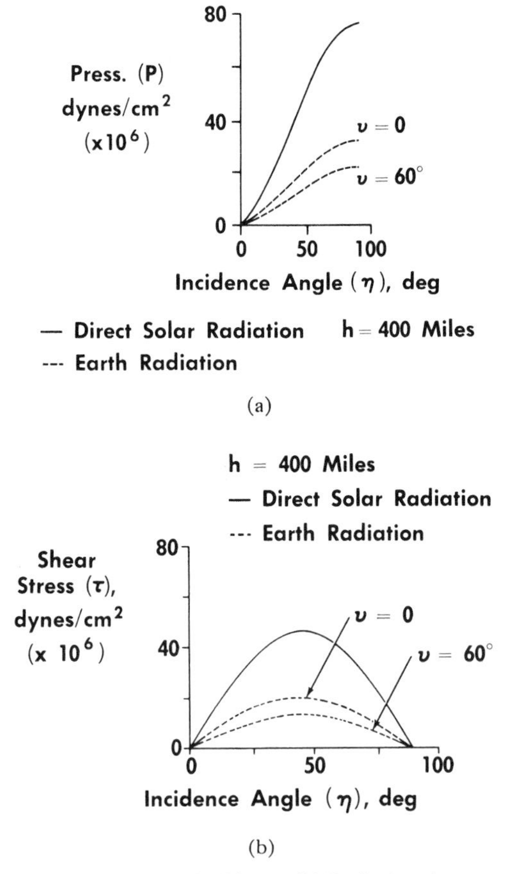

FIG. 6. (a) Radiation pressure vs. incidence. (b) Radiation shear stress vs. incidence.

sun and the earth itself. Figure 5 shows the values, at 400 miles altitude, of the earth-emitted radiation, as well as that reflected from the daylight portion of the earth, plotted against the angular displacement ν of the satellite position vector. The angular displacement is measured from the earth-sun line. The earth's radiation is seen to be considerably less than the solar value, but not usually by an order of magnitude. The resulting pressure and shear stress contributions are shown in Fig. 6, and the radiation effects are seen to be of the same order as the aerodynamic effects.

IV. Shielding

Shielding of one surface by another, from the relative wind and radiant flux, will result in considerable disturbance torques. This is a prime source of fundamental asymmetry, and requires careful analysis. Certainly no disturbance torque analysis can be more accurate than the shielding analysis performed. The geometry of a representative satellite is shown in Fig. 7. In order to analyze the shielding of this geometry, an

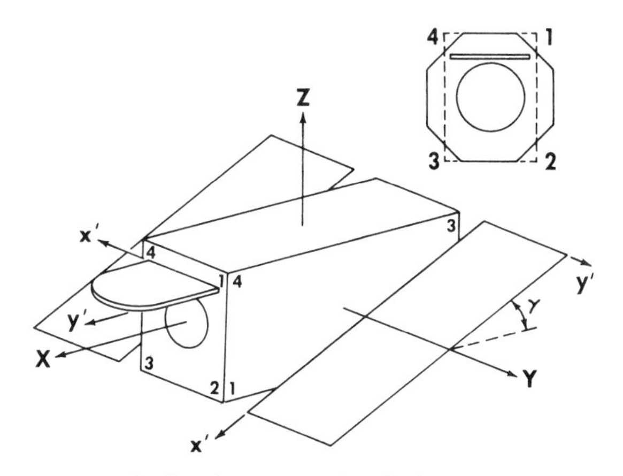

Fig. 7. Reduced geometry for shadow calculations.

incident relative flux vector of any kind is utilized to cast a shadow pattern. The map of this shadow pattern on an intercepting surface is obtained (cf. Figs. 8, 9) and the shadowed or shielded areas and area moments are integrated. The remaining active areas and their moments are then utilized to compute the disturbance torques. This procedure is repeated for each of the relative vectors–aerodynamic and radiation. Programmed on a digital computer, this analysis, along with aerodynamic and radiation analyses, enables one to produce the disturbance torque time histories of Figs. 12-14, for the relative vectors of Figs. 10 and 11.

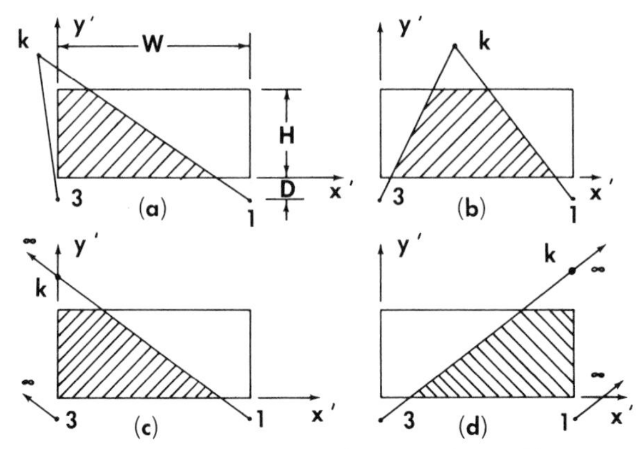

FIG. 8. Geometry of shadow on solar paddles.

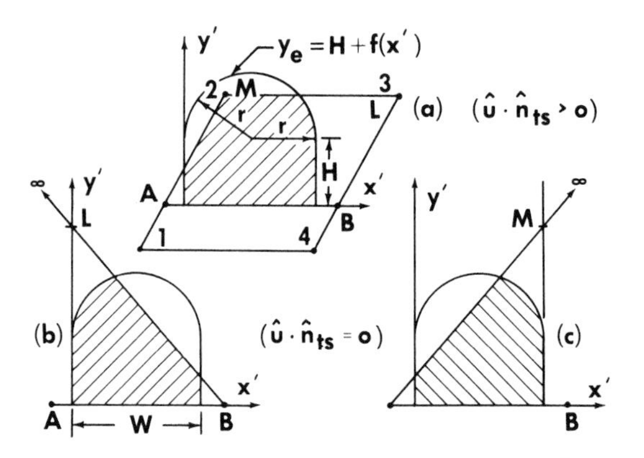

FIG. 9. Geometry of shadow on the sunshade.

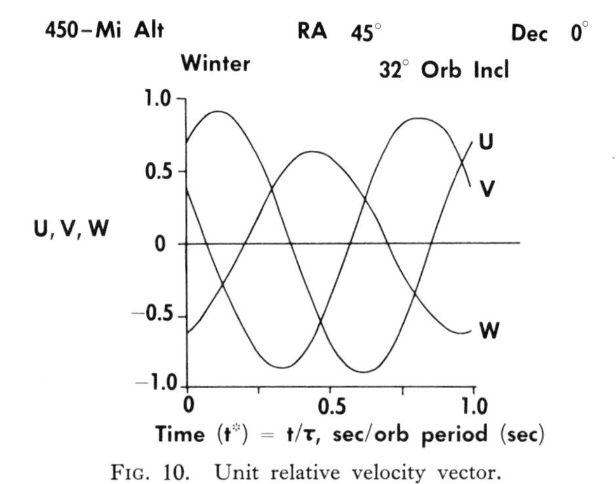

FIG. 10. Unit relative velocity vector.

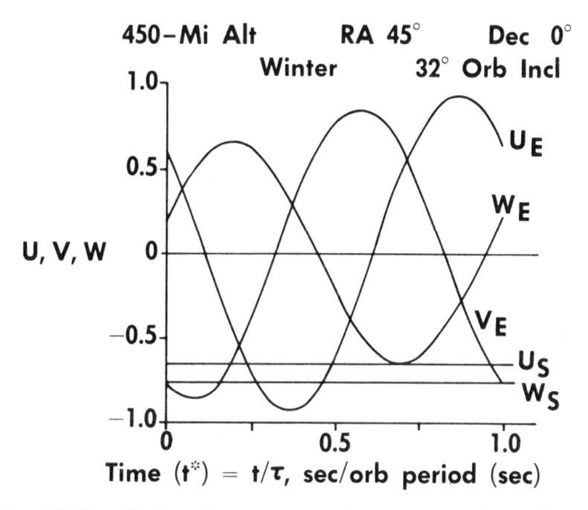

FIG. 11. Unit relative flux vectors for solar and earth radiation.

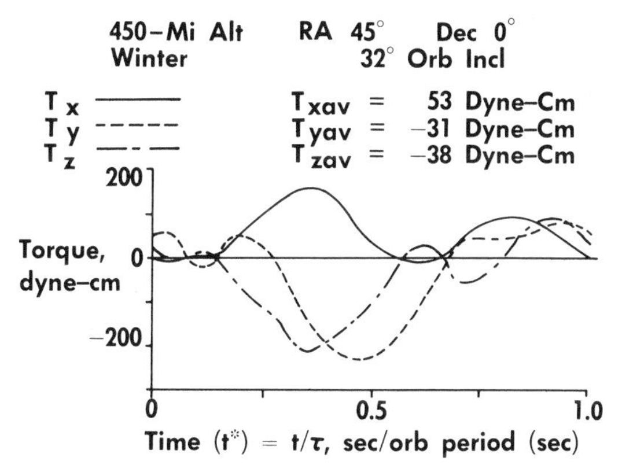

FIG. 12. Aerodynamic torque.

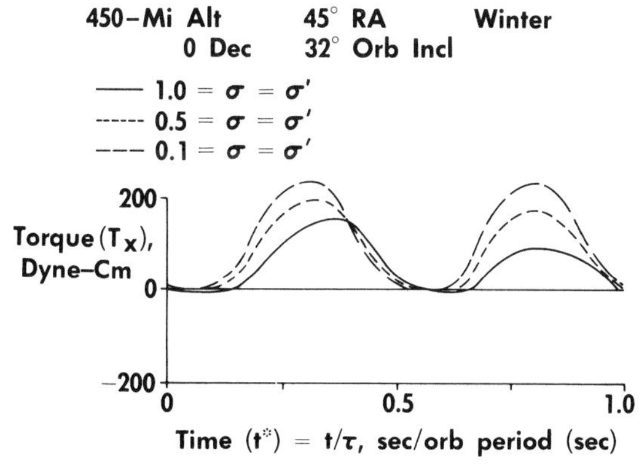

FIG. 13. Aerodynamic torque—effect of accommodation coefficients.

This satellite is constrained to remain in a fixed orientation with respect to the sun and stars in order to perform a successful experiment. In addition, in order that the power derived from the solar paddles be maximized, the relative sun vector is maintained in the satellite's plane of symmetry. Thus, the relative velocity vector and the earth's relative vector revolve about the satellite in their respective planes, while the relative Sun vector is constant and has no side component. Therefore, the Earth radiation and aerodynamic torques tend to oscillate while the one component of solar torque remains constant until it disappears during occultation.

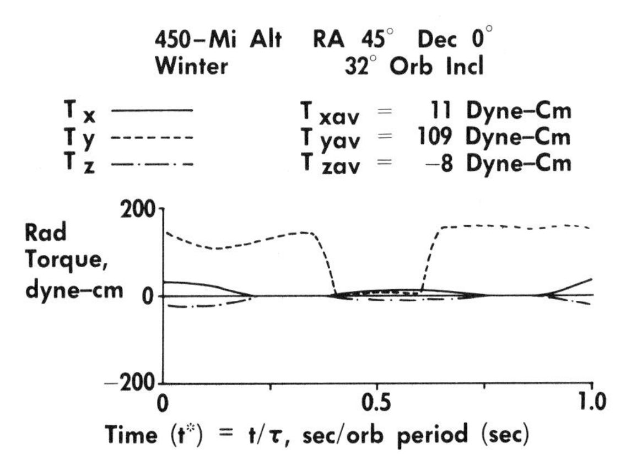

FIG. 14. Radiation torque.

For this orbit, the planes of the rotating relative vectors are skewed, and so do not contain any of the vehicle axes. The effect of shielding then is to introduce asymmetry that does not cancel itself over an orbit, and torques are accumulated about each axis.

V. Conclusion

The disturbance torques due to aerodynamic and radiation pressures and shear stresses have been formulated on the basis of convex geometry. The effects of concave geometry and the variation of the accommodation coefficients have been discussed and weighed. The reflected and emitted radiation from the earth and direct solar radiation have all been included. The effects of shielding have been mentioned and torque time histories presented for an example orbit of a particular satellite.

Appendix. Symbols

c	Velocity of light
E	Internal energy flux (aerodynamic); radiation energy flux (thermal units, cal/cm^2 sec)
f	Molecular distribution function
m	Molecular mass
P	Pressure
R	Gas constant
s	Speed ratio
S	Radiation energy flux (mechanical units, ergs/cm^2 sec)
T	Temperature, absolute
U	Relative velocity of the free stream
u, v, w	Components of molecular velocity
u_0, v_0, w_0	Components of macroscopic velocity of the free stream
x, y, z	Local coordinate system at a surface
α	Thermal accommodation coefficient (aerodynamic); surface absorptivity (radiation)
η	Local incidence measured from the surface
θ	Local incidence measured from the outward normal
ν	Angle between the satellite position vector and the earth-sun line
ρ	Air density; surface reflectivity (radiation)
σ	Surface reflection coefficient for tangential momentum
σ'	Surface reflection coefficient for normal momentum
$\tau, d\tau$	Tangential stress, element of volume ($dx\, dy\, dz$)
τ'	Surface transmissivity (radiation)
$d\omega$	Element of volume in velocity space ($du\, dv\, dw$)

Subscripts

d	Diffusely
E	Earth
e	Emitted
i	Incident
r	Reflected, or re-emitted
s	Specularly
S	Sun
w	Re-emitted in Maxwellian equilibrium with the surface

REFERENCES

1. S. A. Schaaf, and P. L. Chambre, Flow of rarefied gases. "High Speed Aerodynamics and Jet Propulsion" (H. W. Emmons, ed.), Vol. III, Section H. Princeton Univ. Press, Princeton, New Jersey, 1958.
2. S. A. Schaaf, and L. Talbot, Mechanics of rarefied gases. "Handbook of Supersonic Aerodynamics," Vol. 5, Section 16. NAVORD Rept 1488, February 1959.
3. G. N. Patterson, "Molecular Flow of Gases." Wiley, New York, 1956.
4. L. Talbot, Rarefied gasdynamics. *Advances in Appl. Mech. Suppl. I.* Academic Press, 1961.
5. R. A. Minzner, K. S. W. Champion, and H. L. Pond, The ARDC Model Atmosphere 1959, *ARDC TR 59-267*, August 1959.
6. E. C. Cornford, D. G. King-Hele, and R. H. Merson, Recent studies of satellite orbits, *IAS Paper No. 59-141*, October 1959.
7. W. J. Evans, *OAO* Stabilization and Control Subsystem, Phase I, Rept. XA-252 R-01.1, -02.0, Grumman Aircraft Eng. Corp., Bethpage, New York, June 1961.
8. M. Jacob, "Heat Transfer," Vol. I, pp. 23-40. Wiley, New York, 1956.
9. K. A. Ehricke, "Space Flight," pp. 139-148. Van Nostrand, Princeton, New Jersey, 1960.
10. R. L. Grossman, A summary Review of Radiation Pressure on Orbiting Satellites, Grumman Aircraft Eng. Corp., Bethpage, New York, December 1960.
11. A. J. Katz, Determination of Thermal Radiation Incident upon the Surfaces of an Earth Satellite in an Elliptical Orbit, Rept. XP 12.20, Grumman Aircraft Eng. Corp., Bethpage, New York, 1960.
12. OAO Thermal Control Phase I, Rept *XP-252-R-04.0*, Grumman Aircraft Eng. Corp., Bethpage, New York, June 1961.
13. R. E. Stickney, An Experimental Investigation of Free Molecule Momentum Transfer between Gases and Metallic Surfaces, *Univ. of Calif. Tech. Rept. HE-150-182*, January 1962.
14. I. M. Cohen, Free Molecule Flow Over Non-Convex Bodies, *Princeton Univ. Rept. 497, AFOSR TN 60-190*, February 1960.

Forces and Torques Due to Coulomb Interaction with the Magnetosphere

S. FRED SINGER[1]

National Weather Satellite Center, Weather Bureau, Department of Commerce, Washington, D. C.

I. INTRODUCTION

IT IS USUAL TO NEGLECT aerodynamic forces and torques above about 500 km, except for bodies having a very large area-to-mass ratio, e.g., the Echo balloon; even there the effects of solar radiation pressure are considered more important. (See, e.g., reference [1].)

This attitude is not always justified. For example, the writer has concluded that in the case of bodies which are *very small* (in addition to having a high A/M ratio) the effects of the surrounding atmosphere may predominate; the reason is that particles acquire an electric charge so that the Coulomb field presents an effective area which is very much larger than the geometric cross section [2].

Similar ideas have been applied to large satellites by Lehnert, Jastrow and Pearse, Kraus and Watson, Beard and Johnson, and Wyatt. The consensus favors only a very small charge drag effect, although opinions are quite divided. (For a review, see reference [3].)

In the case of small particles, such as interplanetary dust [4], or the West Ford needles [5], the Coulomb drag may play a most important

[1] On leave: University of Maryland, College Park, Maryland.

role in relation to other perturbing forces such as radiation pressure, magnetic (Lorentz) forces, etc.

(It is important to realize that nearly all of space is pervaded by ionized gas, mostly a hydrogen plasma. Above a couple hundred kilometers, i.e., above the ionosphere, the earth's atmosphere can be considered as a plasma. The interplanetary gas is almost completely ionized and has a density of 20-50 ions/cm^3 near the earth's orbit.)

What we wish to show here is the importance of Coulomb drag for larger bodies, under special circumstances, particularly in the production of torques.

II. The Electric Charge of Space Vehicles

It is a very difficult problem to obtain reliable and universally agreed-to values of vehicle potential. The potential is determined mainly by competition between (i) accretion of negative charges and (ii) the solar photoelectric effect, as pointed out by the writer [2]; (iii) trapped particles in the earth's radiation belts play a less important role.

Below about 1 earth radius (4000 miles) process (i) is likely to predominate; in interplanetary space process (ii) is more important; hence the potential changes from negative to positive values somewhere around 4000 miles [4]; a more detailed treatment is in preparation.

For purposes of the present paper we adopt a potential of -5 volts for near-earth satellites, and a potential of $+30$ volts for interplanetary vehicles.

III. Calculation of Coulomb Force

The difficult problem is the calculation of the force experienced by a charged body moving through a plasma. The drag is proportional to the ion density (and therefore electron density) of the surrounding plasma, but depends in a complicated way on the velocity and charge of the body. The problem has not yet been solved in its complete form. The writer has been able to develop simple expressions which give the drag within a factor 2, and often much closer, for very small particles (see Appendix).

Here we shall be mainly concerned with large drag bodies where the geometric cross section represents a good approximation to the effective drag area.

The important difference from ordinary aerodynamic drag is that in the present case an electrically charged wire net can appear as effective as a solid surface.

Each case needs to be examined separately, but some generalizations are possible:

(1) The degree of transparency of the wire net is a function of the wire spacing (mesh), the wire size, and the electric potential. If the latter is large enough so that the potential at the center of a mesh loop is greater than the kinetic energy of a plasma ion (as viewed from the vehicle), then the transparency is nearly zero. It is seen therefore that, with a fixed mesh geometry and a means of varying the potential, it is possible to adjust the drag force from nearly zero to a maximum value, corresponding to transparency of nearly 100 % (potential equal to plasma potential, i.e. $V = 0$) and 0 %, respectively.

(2) The *direction* of the Coulomb force is determined by the design of the drag body. For a spherical drag body the force is always in the direction of relative velocity between body and plasma. For a flat drag body the effective force is normal to the body's surface; it can therefore be used as a "sail."

IV. Some Vehicle Applications

A. An Echo Communication Satellite

At first glance it may appear desirable to replace the present Echo balloon by a spherical wire net. Its ordinary aerodynamic drag would be reduced by a large factor, and its weight might even be less (depending upon design).

But consideration of Coulomb drag shows that the total drag would remain about the same (particularly if the mesh distance d is less than the Debye length[2] h) [6]. On the other hand, the disturbing effects of solar radiation pressure would be virtually eliminated.

B. Earth Satellite with Large Antenna

Imagine a design for a space vehicle having a large light-weight antenna constructed of wire. Because of interference effects the antenna is placed on the end of a long boom. Some typical values might be: antenna area = 10^5 cm^2; boom length —1000 cm. The Coulomb force on the antenna will be given by Eq. (A5a) of the Appendix. Take an operating

[2] The Debye length h is the conventionally defined screening distance $h = [kT/4\pi n_e e^2]^{1/2} = 6.90\,(T/n_e)^{1/2}$ in centimeters.

altitude of 4000 km, where the H^+ density is $1.5 \times 10^3/\text{cm}^3$, and the satellite velocity $\sim 6 \times 10^5$ cm/sec.

$$F = 1.5 \cdot (10^5\ \text{cm}^3) \cdot (1.5 \times 10^3\ \text{cm}^{-3})\,(1.6 \times 10^{-24}\ \text{gm}) \cdot [(6 \times 10^5)^2 + (6 \times 10^5)^2] \sim 2.6 \times 10^{-4}\ \text{dynes}$$

The torque will be of the order of $2.6 \times 10^{-4} \cdot 10^3 = 2.6 \times 10^{-1}$ dyne-cm.

This value may be compared with a typical gravitational torque. Assume a dumbbell with mass 10 kg divided between the two halves which are effectively separated by a distance of 200 cm. The torque at 4000 km altitude is of order

$$10^4\ \text{gm} \cdot (380\ \text{cm/sec}^2)\,(100\ \text{cm}/1.04 \times 10^9\ \text{cm})\,(100\ \text{cm}) = 3.6 \times 10^1\ \text{dyne-cm}$$

As another illustration consider radiation pressure torques which may present a problem, e.g., in astronomical satellites.

$$\begin{aligned}\text{Torque} &= (\text{solar constant}) \times (\text{speed of light})^{-1} \times (\text{area}) \times (\text{moment arm}) \\ &= (1.4 \times 10^6\ \text{erg/cm}^2\text{-sec})(3 \times 10^{10}\ \text{cm/sec})^{-1}\,(10^5\ \text{cm}^2) \times (100\ \text{cm}) \\ &= 5 \times 10^2\ \text{dyne-cm}\end{aligned}$$

But this value may be reduced (by proper design) to about 1 %, i.e., to 5 dyne-cm. On the other hand, the Coulomb torque would increase when the satellite operates at a lower altitude where the ion density is higher. Therefore the Coulomb torque could produce a disturbing, and generally unanticipated, effect.

V. Control Applications

From the foregoing discussion it should be fairly obvious how the charge drag effects can be used to control the torque acting on space vehicles. Specifically, there are several ways of varying the force: (it may not be practical to vary the moment arm).

(i) Change in potential of drag body to vary the transparency. This can be accomplished by inserting a variable "battery" between the vehicle and the drag body.

(ii) Change in drag body orientation will change the direction of the force (in the case of a nonspherical drag body). One can picture various sail applications, or a propeller application to spin a vehicle about an axis parallel to its velocity vector.

(iii) A spherical drag body will always line up behind the vehicle in a direction along the relative velocity vector between vehicle and plasma.

It should be pointed out explicitly that a certain amount of power is required for maintaining the drag body at a potential different from its equilibrium value. It is necessary to supply a current to balance the current to or from the plasma [7].

VI. Some Scientific Applications

Several research applications can be listed for this technique:

(i) Direction of solar wind. In interplanetary space a drag body can act essentially as a wind vane.

(ii) Direction of vehicle velocity. If the plasma has a negligible streaming speed, the drag body will tend to point in the direction of the vehicle's velocity vector. This application is most suitable for low-altitude earth satellites.

(iii) Measurement of plasma density by observing the torque (suggested to the author by Mr. L. Rosen of Jet Propulsion Laboratory).

(iv) Of first order of importance is a measurement of the parameters of Coulomb drag, such as vehicle potential. An over-all measurement of the drag would be useful. Our suggestion would be to observe the orbit parameters of a radar target, made of wire netting, and placed into orbit.

Appendix. Calculation of the Charged Drag in a Plasma

We wish to find the drag force acting on a body (sometimes referred to as the *test* particle) which moves through a plasma with velocity w. The body carries an electric charge Ze. The drag force is given by the summation of all the Coulomb interactions. Unfortunately, the charge is at least partly determined by the accretion of plasma particles, which in turn depends on their velocity distribution and on w. But most important, the shielding depends on these same quantities in a complicated way, leading to an asymmetric distribution of particle trajectories and to a modified accretion. This complete problem has not yet been solved in a self-consistent manner. In the following treatment we shall assume the charge as given.

We replace the complicated real situation by an idealized model, which, however, should give a closely similar result for the drag. We define an impact parameter p_0 for which the deflection of the ion is 90 deg. For $p < p_0$, we consider the screening negligible, and for $p > p_0$, the screening is taken into account.

The potential up to a distance p_0 is given by the vacuum formula; ϕ_0 is defined as being twice the kinetic energy of the plasma particles (assumed to be singly charged ions or electrons). Hence,

$$e\phi_0 = m_1(w^2 + \alpha C_1^2) \tag{A1}$$

Here m is the mass, C_1 the rms velocity of the field particles; α is of order unity. For a cold plasma, or for heavy ions, w, the velocity of the body, determines the kinetic energy.

If the body is a sphere of radius a and potential V, then

$$p_0 = aV/\phi_0 = aVe[m_1(w^2 + C_1^2]^{-1} \tag{A2}$$

The drag exerted on the particle for impact parameters $p < p_0$ can be viewed in terms of a drag coefficient C_d of ~ 1.5, since the ion deflection will always exceed 90 deg. The (partial) drag force can therefore be written as

$$F_0 = \tfrac{1}{2} C_d n_1 m_1(w^2 + \alpha C_1^2)\, \pi p_0^2 \tag{A3}$$

To obtain the contribution to the drag from $p > p_0$ we proceed as follows. We regard the potential as screened beyond p_0 within a distance of scale h. The Debye length (see, e.g., reference [6]) is the conventionally defined screening distance

$$h = [kT/4\pi n_e\, e^2]^{1/2} = 6.90\,(T/n_e)^{1/2} \qquad \text{in centimeters}$$

The *additional* drag then is given by adding to the "hard" area (πp_0) the area $(2\pi p_0 h)$ which, however, is reduced by the factor $\ln \Lambda = \ln (1 + h/p_0)$ to take account of the fact that the deflections are now less than 90 deg. We shall, therefore, use an effective area

$$A_{\text{eff}} = \pi p_0^2 + 2\pi p_0\, h \ln \Lambda \tag{A4}$$

Our aim is to get a drag expression good within a factor of 2, which is not much worse than the uncertainties in expressions for neutral drag. Hence,

$$F_c \sim A_{\text{eff}}\, n_1 m_1(W^2 + C_1^2) \qquad w > C_1 \tag{A5a}$$

$$F_c \sim A_{\text{eff}}\, n_1 m_1\, w\, C_1 \qquad w < C_1 \tag{A5b}$$

We want to find the drag force on a large spherical body of radius a. If the body is solid and its charge zero, then $A_{\text{eff}} = \pi a^2$ and the problem presents no special features.

If the body is not solid, but made out of a fine wire net, the size of the body's potential becomes important. Assuming the body to be a satellite ($w >$ ion thermal velocity), then a positive body (with wire mesh spacing $<$ Debye length h) will present an essentially solid surface to positive ions, provided the potential at the center of a mesh square is greater than the ion kinetic energy. For smaller potentials, the grid will be translucent, and for zero potential very nearly transparent.

A negatively charged grid will experience almost exactly the same drag (as can be verified by drawing equipotential curves and looking at ion trajectories). In a detailed evaluation the size of the mesh and the wire size are of importance.

References

1. I. I. Shapiro and H. M. Jones, *Science* **134**, 973 (1961).
2. S. F. Singer, *in* "Scientific Uses of Earth Satellites." Univ. of Michigan Press, Ann Arbor, 1956.
3. K. P. Chopra, *Rev. Mod. Phys.* **33**, 153 (1961).
4. S. F. Singer, *Am. Astronaut. Soc. Lunar Flight Symp. December 1960*, Plenum Press, New York, 1962.
5. S. F. Singer, *Nature* **192**, 303 (1961); 1061 (1961).
6. L. Spitzer, "Physics of Fully Ionized Gases." Interscience, New York, 1956.
7. D. B. Beard and F. S. Johnson, *J. Geophys. Res.* **66**, 4113 (1961).

Dynamical Considerations Relating to the West Ford Experiment

R. A. LYTTLETON[1] and S. FRED SINGER[2]

Jet Propulsion Laboratory,
Pasadena, California

I. INTRODUCTION

THIS PAPER HAS BEEN motivated by the failure to detect the West Ford needles, launched on October 21, 1961, and the fact that as yet no explanation for this seems to have been found. The suggestions herein put forward may only provide a partial explanation for the failure, and in any case the processes proposed involve factors difficult to assess. But the general lines of the mechanism seem sufficiently definite that some further study may be considered desirable to investigate their relevance to the West Ford project and any further similar attempts.

The work of this paper was communicated to Dr E. R. Dyer, Secretary of the West Ford Project of the Space Science Board, in March 1962, at his request, and we understand that modifications in the design of subsequent West Ford packages have been made that meet the dynamical objections explained below.

II. OUTLINE OF EFFECTS INVOLVED

We suggest that the following three main effects may have operated to prevent the system functioning in the manner intended:

[1] Present address: St. Johns College, Cambridge, England.
[2] Present address: National Weather Satellite Center, U. S. Weather Bureau, Washington, D.C.

(i) A change in the axis of rotation of the needle dispenser brought about by dissipation of dynamical energy of rotation through internal imperfections of elasticity.

(ii) A consequent much-decreased dispensing rate of the needles owing to the reduced angular velocity, and to possible interference effects associated with the design of the package.

(iii) A rapid subsequent spatial dispersion of such needles as may have been released through the action of Coulomb electrical drag [1].

III. CHANGE OF ROTATION AXIS OF THE PACKAGE AS A WHOLE

We understand that the package had weight about 76 lb. with the general shape of a right circular solid cylinder of over-all length 16 inches and diameter of cross-section $5\frac{1}{2}$ inches, and contained about 350×10^6 copper needles embedded in naphthalene. The needle-dispensing container has been described as a supporting spool consisting of two heavy steel end-plates of diameter $5\frac{1}{2}$ inches mounted transversely to an aluminum tube 16 inches long, around which the needles were mounted symmetrically in "pineapple" slices each 0.7 inch high and presumably $5\frac{1}{2}$ inches in diameter (with a small central region cut out where the tube passes).[3] The needles were thus set in the package parallel to the axis of the cylinder.

The objective was to launch the system in such a way that it would spin with angular velocity of a few revolutions per second about the axis of symmetry, which as we shall show would have been the axis of smallest moment of inertia. Denoting the length of the cylinder by L, and its outer radius by R, and supposing the system to have uniform density ρ, then for the principal moments of inertia, I_1, I_2, I_3 (about the axis), we have approximately

$$I_1 = I_2 = \tfrac{1}{4}M(R^2 + L_3^2), \qquad I_3 = \tfrac{1}{2}MR^2 \tag{1}$$

since the effect of the hollow at the axis (if present) can safely be ignored. Thus we have, with the above numerical values,

$$I_1 : I_3 = 6.14 : 1 \tag{2}$$

and the intended axis of rotation is therefore that of smallest moment of inertia by a considerable factor.

[3] All values given here are approximate only, since detailed data have not yet been available to the writers.

Now it is a well-known result in dynamics that a perfectly rigid body is rotationally stable when spinning about either its axis of greatest inertia or its axis of least inertia. Both these motions are "ordinarily" stable, that is, if the system is slightly disturbed, the motion remains in the immediate neighborhood of the original motion, it being supposed that there are no sources of dissipation of energy. But for a given angular momentum, rotation about the axis of greatest inertia involves less energy than rotation about the axis of least inertia. For if this angular momentum is h, and we denote the respective angular velocities by w_1 and w_3, then

$$h = I_1 w_1 = I_3 w_3 \tag{3}$$

while the corresponding kinetic energies would be

$$T_1 = \tfrac{1}{2} I_1 w_1^2 = h^2/2I_1 \tag{4}$$

$$T_3 = \tfrac{1}{2} I_3 w_3^2 = h^2/2I_3 \tag{5}$$

Thus we have

$$T_1 : T_3 = I_3 : I_1 = 1 : 6.14 \tag{6}$$

so that for the same angular momentum, which cannot be altered in an isolated system, rotation about the axis of greatest moment represents the state of absolute minimum energy. (See Fig. 1.)

Accordingly, if there were any way that dissipation of dynamical energy could occur within the system, rotation about the axis of least inertia would be secularly unstable, and the system if so started would (under the slightest disturbance) gradually depart from it more and more and finally come to rotate about the axis of greatest moment. The motion at any intermediate stage would be (almost) one of Eulerian nutation, with a finite angle between the axis of symmetry and the angular momentum vector h, and it is the accelerations associated with such motion that would lead to inernal strains and thence to dissipation of energy through imperfections of elasticity. For the needle package, the initial rotation axis can scarcely have coincided accurately with the axis of symmetry, and also there would certainly be some nonrigidity of the package.

If we estimate the initial rotation speed to have been about 4 revolutions a second, so that $w_3 = 8\pi$ radians sec^{-1}, the initial energy would have been about 3×10^4 ergs per gram on the average through the package. This is quite small compared with the heat capacity of copper, for example, which is 3.8×10^6 erg gm^{-1} deg^{-1} near 0°C, and about half this at -200°C. Thus there would be no difficulty about accommodating dissipated energy and conducting it to the outer surface or to the naphthalene. The total initial energy of rotation would have been about 10^9 ergs.

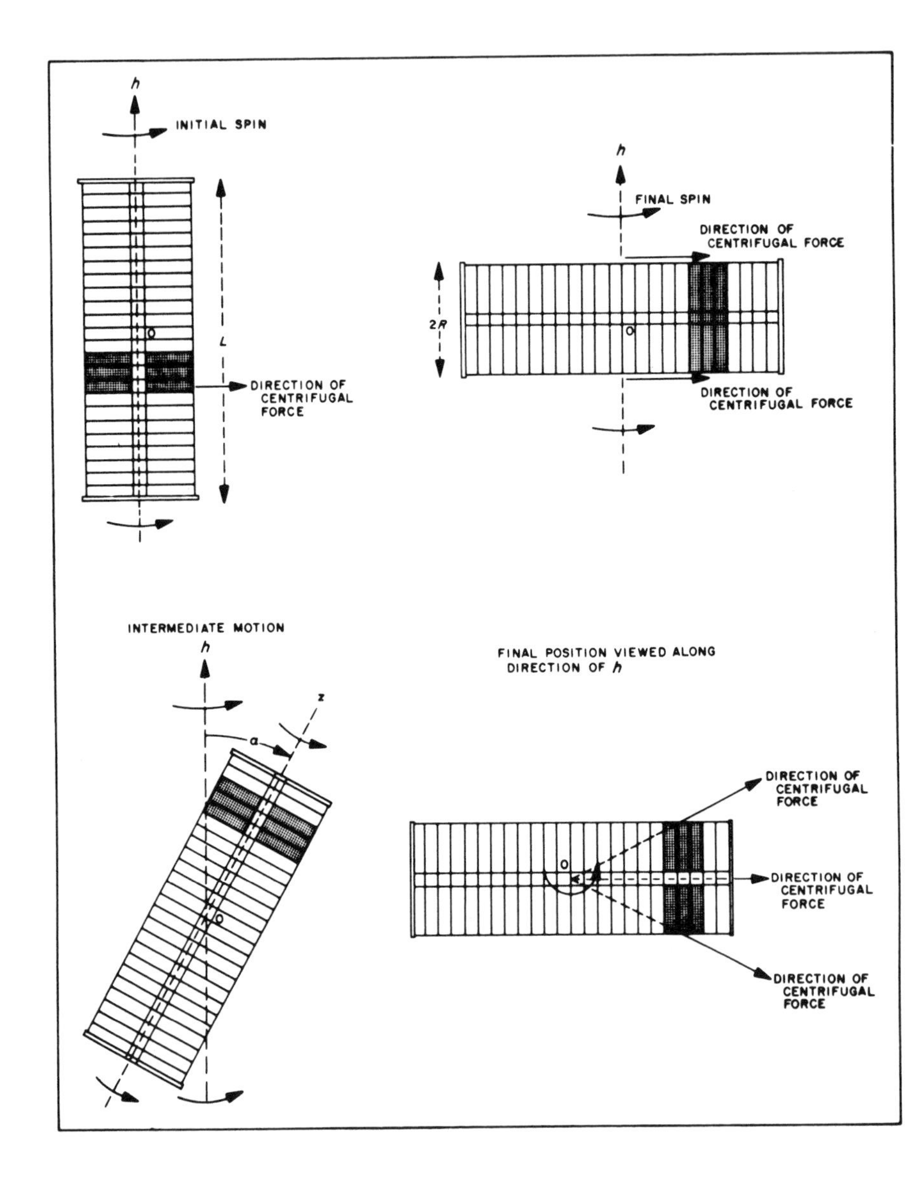

Fig. 1.

IV. Analysis of Needle Dispensing

Presumably the needles are meant to be removed by centrifugal force from the surface of the package as the naphthalene evaporates. In the initial position, the acceleration available for this would be $w_3^2R = 4.4 \times 10^3$ cm sec^{-2}, and each needle would be acted on by a force about 4.5 times its weight. This would have to be capable of overcoming any residual cohesive or frictional forces. It seems likely that a needle would not detach everywhere simultaneously and move off perfectly laterally but rather remain attached at one end or the other so that it is more or less prized off.

But, in the final position, the angular velocity is reduced to $w_1 = w_3/6.14$, and the centrifugal force depends on the point of the surface taken. In the plane through h and the axis of the cylinder, it is always along a generator and increases from zero at the center to a maximum of $\frac{1}{2}w_1^2L = 3.4 \times 10^2$ cm sec^{-2} at the ends—less than a tenth the value in the initial position. For points on the surface in the plane perpendicular to h and through the axis of the cylinder, the centrifugal acceleration has a component normally outward of amount $w_1^2R = 1.2 \times 10^2$ cm sec^{-2}, and a component along the surface increasing from zero at the middle section up to $\frac{1}{2}w_1^2L$, as before, at the ends (see Fig. 1).

There is therefore a very great reduction in the normal force by a factor not greater than about 1/37 as compared with the force available in the initial position. Except near the middle portion of the cylinder, the main effect in the final position would be to slide the needles along the cylinder toward the end-plates, which would hinder their dispersal into space. Systematic motion of the needles toward the ends without release would tend to increase the moment of inertia and lower still further the available centrifugal force.

These simple considerations suggest that the change of rotation axis may seriously weaken the efficacy of centrifugal force for dispersing the needles, and might even prevent dispersal altogether. It would obviously be desirable to make laboratory tests of the release mechanism to ascertain what rotation rates would be adequate.

V. Elastic Dissipation

The point of greatest interest but the one most difficult to settle is the period of time it would take for the package to transfer from its initial state spinning about the axis of least moment to the final state spinning about the axis of greatest moment. This depends on the rate of dissipation at intermediate stages, and the first step toward calculating this would be to determine the (centrifugal) forces to which the elements of the package are subject at any instant. Expressions for the accelerations are readily found by evaluating the vector acceleration

$$\mathbf{a} = \dot{\mathbf{w}} \times \mathbf{r} + \mathbf{w} \times (\mathbf{w} \times \mathbf{r})$$

where $\mathbf{w}$ is the angular velocity of the body and $\mathbf{r}$ the position vector of an element relative to the center of mass. It is found in this way that the impressed forces have a period of approximately $2\pi/w_3$, while the strength of the accelerations is of order $(w_3/6.14)^2$ times distance, which we can denote by μ times distance.

If now E is Young's modulus for the material of the package and ρ its average density, then Q, the work done in an elastic deformation, is found to be

$$\frac{2}{15} A\rho^2\mu^2(L/2)^5E^{-1} = \frac{1}{240} M\rho L^4E^{-1}(w_3/6.14)^4$$

and this is about $5.6 \times 10^{11}E^{-1}$ ergs. The period of the force is approximately $2\pi/w_3 = \frac{1}{4}$, and if f is the proportion of this elastic energy dissipated per cycle, then rotational kinetic energy will diminish at rate

$$\frac{dT}{dt} = -2.3 \times 10^{12} f E^{-1} \text{ erg sec}^{-1}$$

To proceed further, some estimates must be made for f and E. For different substances f would appear to range from about 10^{-2} for solid metals up to somewhere near unity for those of great plasticity. Naphthalene apparently is a substance resembling the tar or asphalt used in street paving [2], so that f may be quite near to 1. As for E, detailed information on the West Ford package has not been available to the writers, but possibly for naphthalene a figure somewhat lower than 10^7 dyne cm^{-2}, which is the value of ordinary rubber, may well be appropriate.

The following table indicates the corresponding estimates (in seconds of time) for various values of f and E:

	E (dyne cm^{-2})		
f	2×10^6	2×10^7	2×10^8
1.0	10^3	10^4	10^5
0.1	10^4	10^5	10^6
0.01	10^5	10^6	10^7

As there are some eight separate cyclic components involved in the acceleration to which the body would be subject, each producing its own contribution to the dissipation, it is entirely possible that this rate would be increased about tenfold. Also, a rotation rate of 7 revolutions per second has now been stated to have been the actual value rather than 4 revolutions per second as adopted here. This would increase the dissipation rate by a factor of about 3 as compared with the kinetic energy. It may thus be expected that the whole energy could be dissipated in a time of the order of 10^3 sec or even less.

It is of interest to compare our result with a quite independent calculation [3]. The author calculates the change in attitude angle for a metal body, a spool whose dimensions are roughly those of the needle dispenser: $I_1/I_3 \sim 0.1$, $E = 2 \times 10^{12}$ dyne/cm², $f = 0.1$, $\omega_0 = 3.2$ rps. He obtains a time of only 3×10^4 sec, considering only one particular mode of dissipation.

VI. Analysis of Needle Dispersion

One may deduce the minimum density of needles necessary to produce a radar echo. From published data the number of needles is 3.5×10^8; the volume into which they spread after one year is $\sim 10^9$ km³, considering only the initial dispersion of velocities and differential radiation pressure effects [4]. Therefore the minimum density required is about 0.5 per km³.

The total number of needles necessary to give a radar return is estimated as follows: The West Ford antenna is 60 ft in diameter and operates at 8000 mc. The beam width therefore is $\sim 2 \times 10^{-3}$ rad. At 5000-km range it covers an area of 10×10 km which contains ~ 50 needles.

The time of dispensing is estimated at 4 days; hence the anticipated rate of dispensing was 1000 per second, or 10^7 per orbital period $T(= 10^4$ sec).

Now we want to consider the effects of Coulomb drag. The rate of shrinkage of the orbit is given as

$$da/dt = \pi^{-1} T F_D/m$$

where the drag force F_D is given [1]

$$F_D \sim 10^{-23} n Z^2 \text{ dyne}$$

or for the West Ford needles:

$$F_D \sim 7 \times 10^{-12} n V^2 \text{ dyne}$$

where n is the plasma density per cm^3 ($\sim 3 \times 10^3$ at 3800 km)
Z the charge in number of electrons
V the potential in volts (~ -3.6 volts)

Substituting numerical values gives

$$da/dt = \pi^{-1}\, 10^4(7 \times 10^{-12} \times 3 \times 10^3 V^2)/10^{-4} \sim 0.7\ V^2 \text{ cm/sec}$$

Therefore,

$$\Delta a/\text{orbit} = (0.7V^2)\, T \sim 9 \times 10^4 \text{ cm/orbit}$$

The rate of change of orbital period is computed as

$$dT/dt = d/dt(ka^{3/2}) = \tfrac{3}{2} ka^{1/2}(da/dt) = 1.5(T/a)\,(da/dt) \sim 1.4 \times 10^{-4}$$

Multiplying by the length of 1 orbit ($= 2\pi a$) gives dispersion/orbit

$$\Delta S = 3\pi T(0.7V^2) \sim 9 \text{ km per orbit}$$

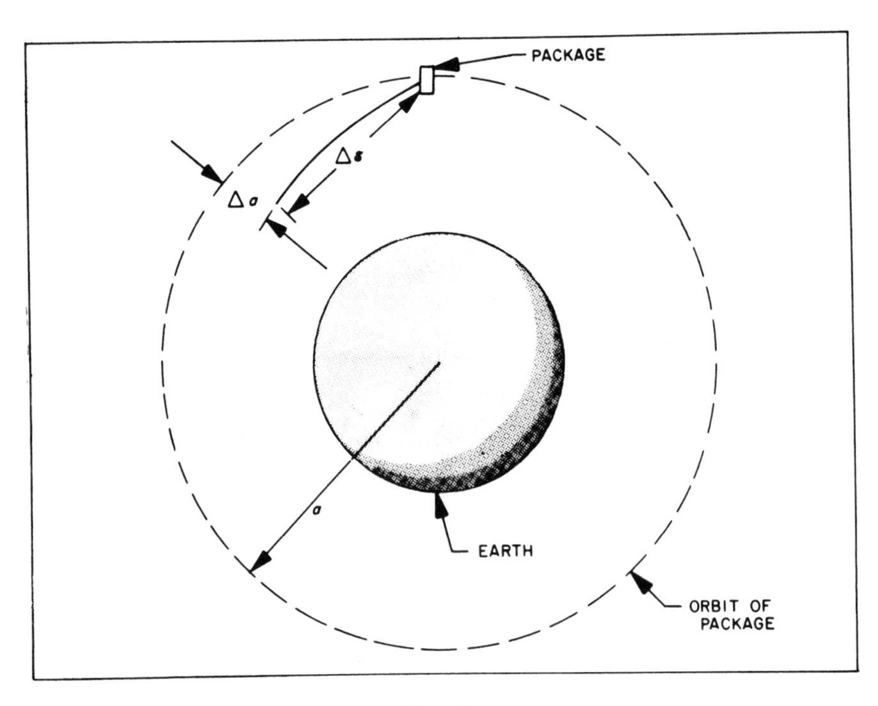

FIG. 2.

The dispersion of needles due to drag in shown graphically in Fig. 2 where ΔS and Δa are indicated. Note that in the absence of drag the needles would simply form a dense cloud around the dispensing package.

The rate of increase of volume occupied by the dispersed needles can be estimated as follows: ΔS increases roughly at $\rightarrow$10 km per 10^4 sec.

Since the radar width amounts to about 10 km we can neglect the diffusion perpendicular to ΔS.

Assuming a radar sensitivity limit of 50 needles (as derived earlier), we conclude that a minimum release rate of 50 needles per 10^4 sec is required to produce and maintain a radar echo.

Putting this conclusion in a different form: If the release rate of needles is less than 50 per 10^4 sec (perhaps 500 would be more conservative), then no echo is maintained because the rate of dispersion of needles exceeds their rate of release.

References

1. S. F. Singer, *Nature* **192**, 303, 1061 (1961).
2. R. H. Wilson, Jr., *Am. Astron. Soc. Preprint 62-54*, March 1962.
3. L. Meirovitch. *J. Astronaut. Sci.* **8**, 110 (1961).
4. I. I. Shapiro and H. M. Jones, *Science* **134**, 973 (1961).

Exploitation of Magnetic Torques on Satellites

RAYMOND H. WILSON, JR.
NASA Goddard Space Flight Center, Greenbelt, Maryland

I. INTRODUCTION; SCIENTIFIC TERMINOLOGY

THE SUBJECT USUALLY CALLED "attitude control," perhaps because it is beset with many nasty problems, may have been such an unwelcome brainchild that it seems to me even to have been misnamed. I shall not mention the possible objection from politically oriented quarters that this volume would be expected to discuss brain-washing techniques, except to say that, until recently, "Webster's New Collegiate Dictionary" recognized only that meaning of "attitude," which we don't intend, namely, the personal and psychological. The transfer to technical use in ballistics and aeronautics to describe orientation in a horizon system of elongated bodies such as a shell or aircraft fuselage may have been natural, since these bodies sometimes seem to act as though they were alive. However, further transfer to space technology is confusing. If the satellite is centrally symmetric and not rotating about a geometric axis, the expression "attitude" is meaningless. Also, since in the weightless, almost frictionless, environment of outer space *all* components of rotation, both long and short period, affect torques and consequent rotational change, the term "rotation control" would seem most accurate and comprehensive. A simpler equivalent term applied to the analogous problem of orienting astronomical telescopes is simply "steering."

II. Magnetic Torque Studies without Satellites

The unexpected discovery in 1825 by the astronomer Arago that there is torque on a conducting body rotating in a magnetic field, even though the conductor is of nonmagnetic material, led Faraday directly to inventions of the electric dynamo and motor [1]. After this, magnetic torques of the Arago type were of little interest, as such, to electrical technology, except as eddy-current and—in ferromagnetic material—hysteresis losses of energy, to be minimized by devices such as insular lamination of the material.

However, the physicist Hertz, about 1890, made a thorough theoretical study of magnetic torques on a conducting sphere by deriving exact mathematical expressions for the induced currents and their resulting energy losses [2]. Later investigators, using the same approach, obtained formulas for magnetic torques on cylinders, but only for rotation about the geometric axis [3].

This was the state of knowledge of magnetic damping torques, when, about 7 years ago, there developed a concrete prospect of a conducting body, an artificial satellite, to be rotating freely in the magnetic fields of outer space. Before Sputnik I, there were theoretical studies applying previous knowledge especially to planned satellites. Two of these by E. Stuhlinger and J. Hooper of the Army Ballistic Missile Agency were especially notable in being followed up by their colleagues Jürgensen, Sanduleak, and Teuber in their report on laboratory checking of the formulas expressing magnetic torques, both on spheres and cylinders [4-6].A most thorough theoretical study by J. Vinti of Ballistic Research Laboratories included the new development of exact expressions for spin-axis wanderings due to magnetic torques resulting from orbital motion of a spherical satellite around the earth's dipole field [7].

III. Magnetic Torque Studies on Satellites

Then came actual satellites in orbits, but since most of them had been planned without consideration of that purpose, studies of magnetic torque effects on them were beset with many confusing difficulties. For example, owing to the principle of least action, a disturbed rotating body slips soon to the position of spinning about the axis of maximum moment, so that long cylinders spin about transverse axes, and aerodynamic torques may confuse magnetic torque effects [8]. Since their approximately spherical shape and high orbits avoided most of the latter difficulty, the first two Vanguard satellites were used successfully by the

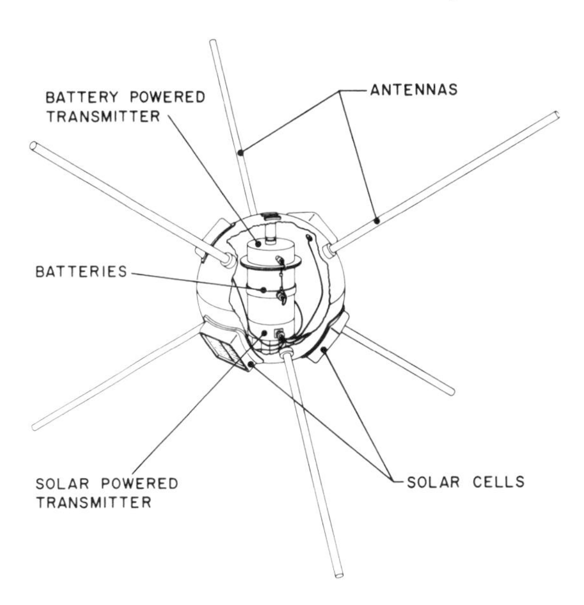

FIG. 1. Vanguard I Satellite.

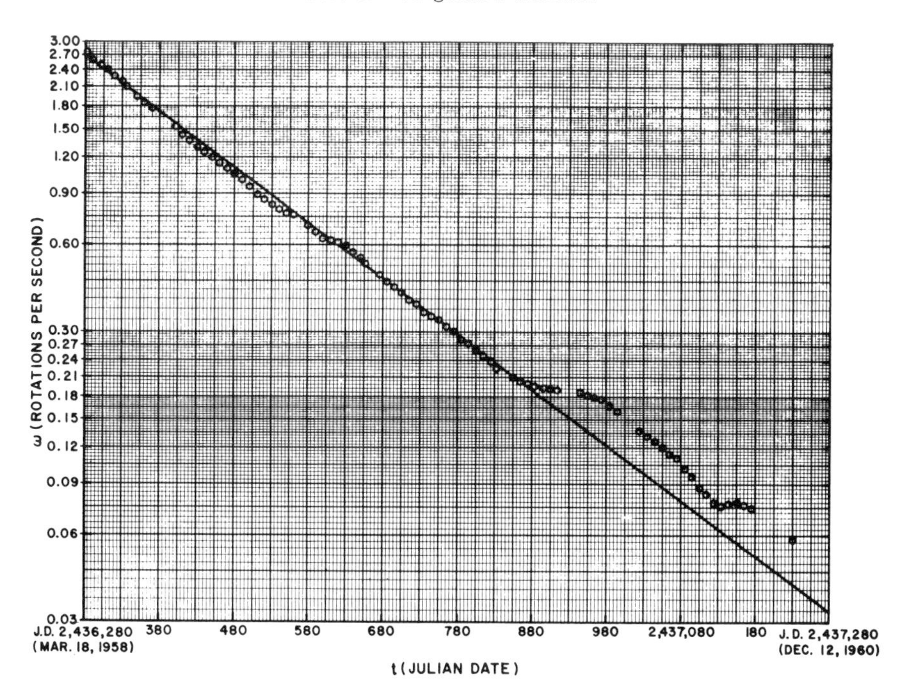

FIG. 2. Observed angular velocity vs. time for Vanguard I (1958 Beta 2). The straight line represents exponential decay with relaxation time of 230 days.

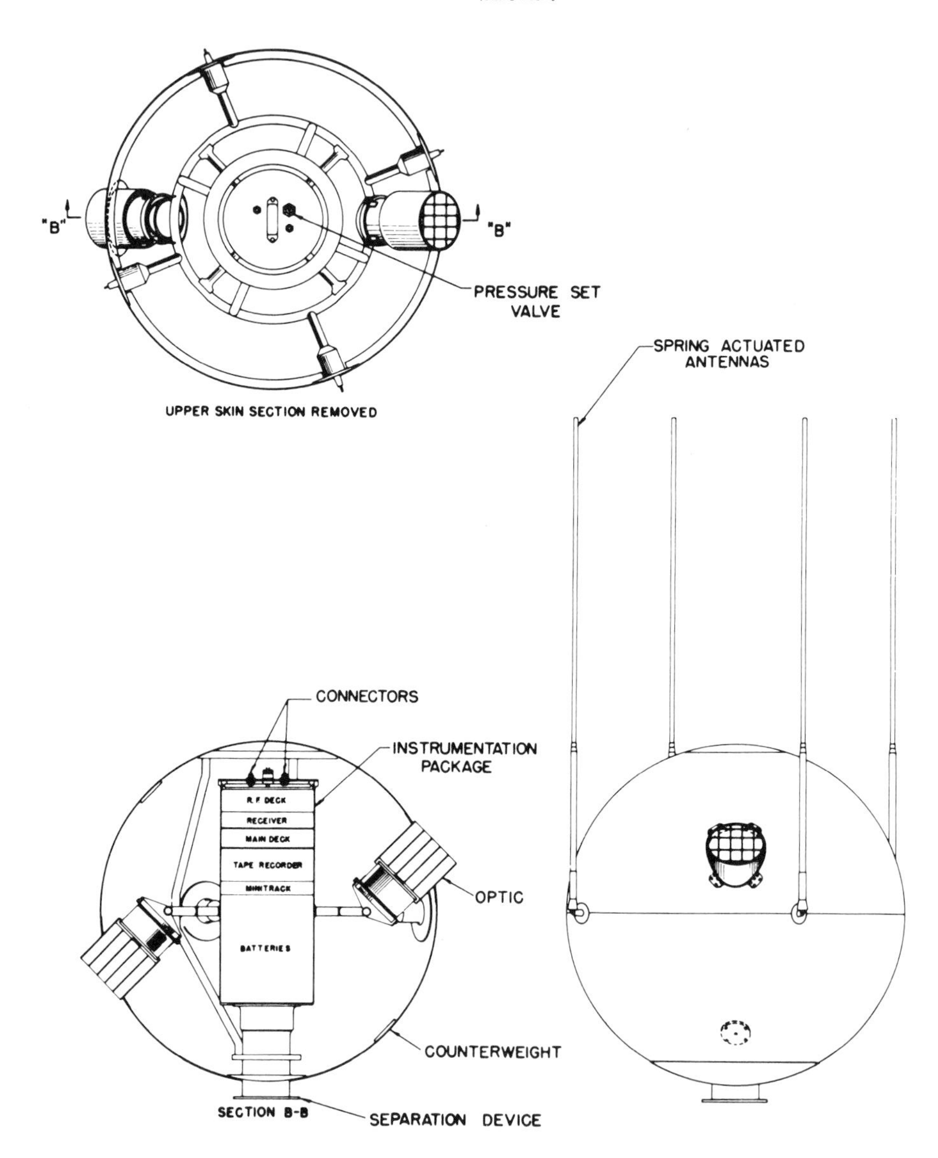

FIG. 3. Vanguard II satellite.

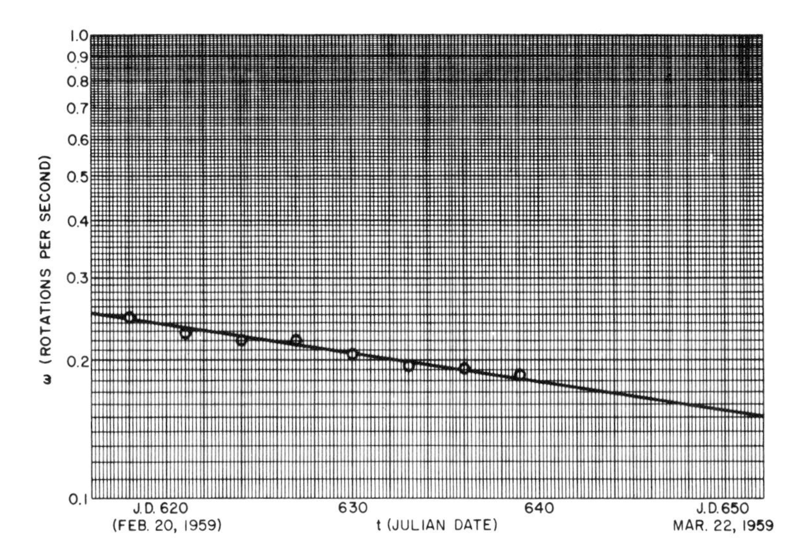

FIG. 4. Observed spin-rate vs. time for Vanguard II (1959 Alpha I). The straight line represents exponential decay with relaxation time of 72 days.

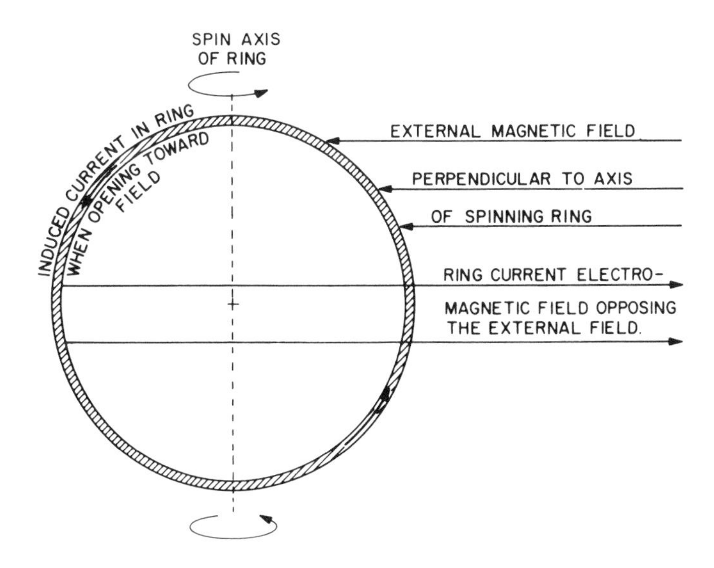

FIG. 5. Interaction of magnetic fields in a spinning, conducting ring. The ring need not be circular.

present author to study the geomagnetic field—a scientific exploitation of magnetic torques [9-11] (Figs. 1-4). However, even in these spherical Vanguard satellites, some parts were cylinders rotating about transverse axes; so it was necessary to derive a new formula for the magnetic torque in this case. This was done by a new approach to the whole problem which expresses the torque on any finite shape as an integral of magnetic torques on elemental infinitesimal rings, so that even though an analytical formula is impossible, a numerical integration for the torque on any shape could be carried out (Fig. 5 and 6).

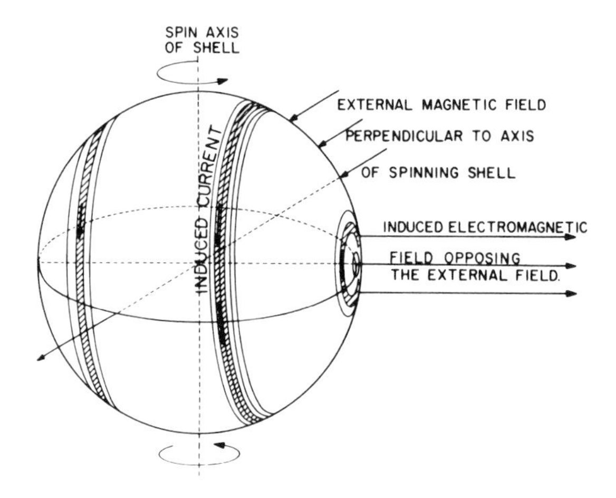

FIG. 6. A spinning spherical shell as an array of elemental rings in a magnetic field.

Certain parts of Vanguards I and II seemed to be affected by relatively high magnetic torques owing to their long cylindrical shapes and/or to very high magnetic permeability of their material. This fact suggested possible technical exploitation of magnetic torques especially for supplying the damping torque always necessary for stable steering, i.e., spin and attitude control [12]. In addition, a magnetic dipole in a satellite could supply a large magnetic restoring torque, although no damping torque could be obtained from a fixed dipole.

IV. ACTUAL AND POSSIBLE EXPLOITATION ON SATELLITES

Actual technical exploitation of magnetic torques has been carried out in the Tiros satellites II and III (dipole restoring torque only) [13], and in the Transit satellites I-B and II-A (both restoring and damping torques) [14]. In planning and discussing the Transit satellites, R. Fischell has emphasized the importance of magnetic hysteresis damping torques on long cylinders made of certain highly permeable magnetic materials (Fig. 7). Evidence of such hysteresis damping combined with the more usual eddy-current damping appeared also from very accurate sun sensory data in the present author's analysis of spin decay of satellite Solar Radiation I, which was launched in attachment to Transit II-A (Figs. 8 and 9) [15].

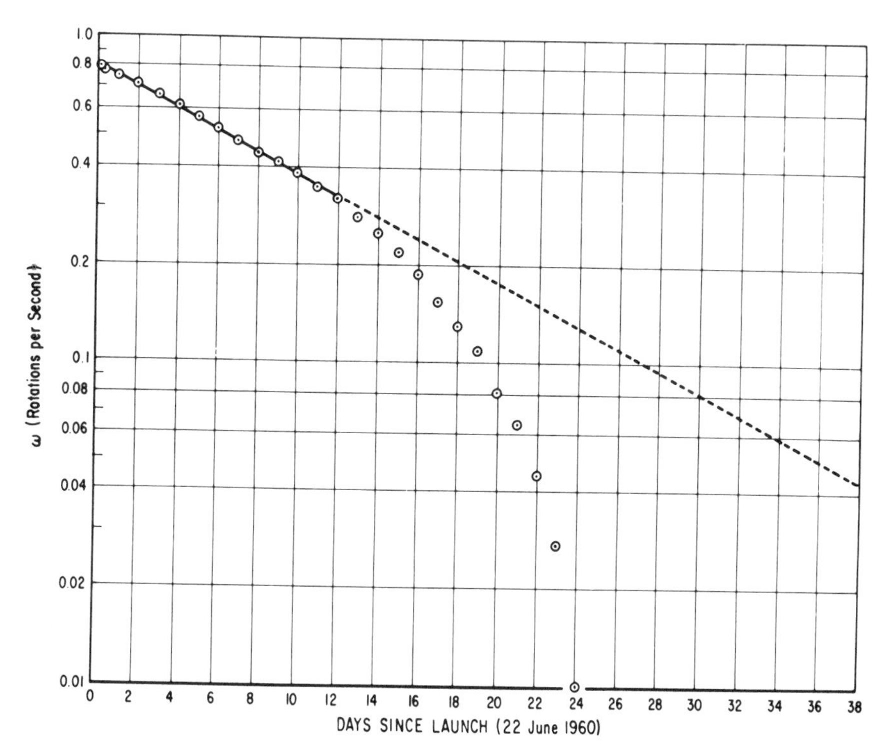

FIG. 7. Observed spin-rate vs. time for Transit 2A (1960 Eta 1), for which hysteresis is dominant cause of decay. Straight line, exponential decay with relaxation time of 13 days; downward curvature of data points shows additional dominant linear decay. (Data from reference [14].)

Future progress in both scientific and technical exploitation of magnetic torques needs to be based on and checked by more accurate and complete rotational studies. Satellite design for exploitation of magnetic torques in any desired form of rotational control may be improved by preliminary theoretical studies, and then checked before launching by measuring static torques on the whole satellite in a rotating magnetic field. A satellite so calibrated could be used as both a magnetometer and thermometer at any point of its orbit with an accuracy limited only by the precision with which its rotational vector could be

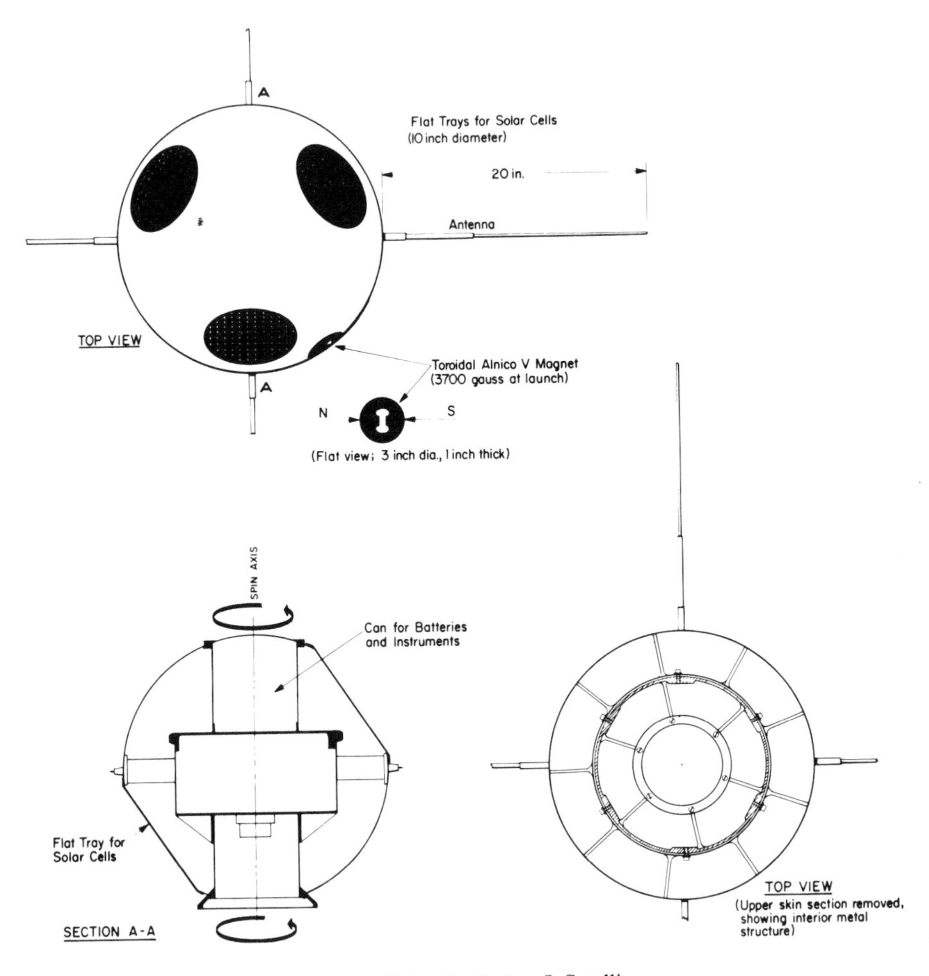

FIG. 8. Solar Radiation I Satellite.

followed. To get sufficiently accurate observations of rotation either there should be telemetry from the satellite of sun or star sensor pulses, or the satellite should have an external reflecting shape so designed that variations in its optical image could be precisely interpreted as rotation vector data [16]. This latter approach can be used even on a radio-dead satellite. Such approach has already been used by the Russian scientist V. N. Grigorevskii in an attempt to correlate sudden spin-rate variations with sudden variations of solar activity [17]. Satellite temperature affects its conductivity and magnetic torque; it, also, can therefore be measured by rotational changes.

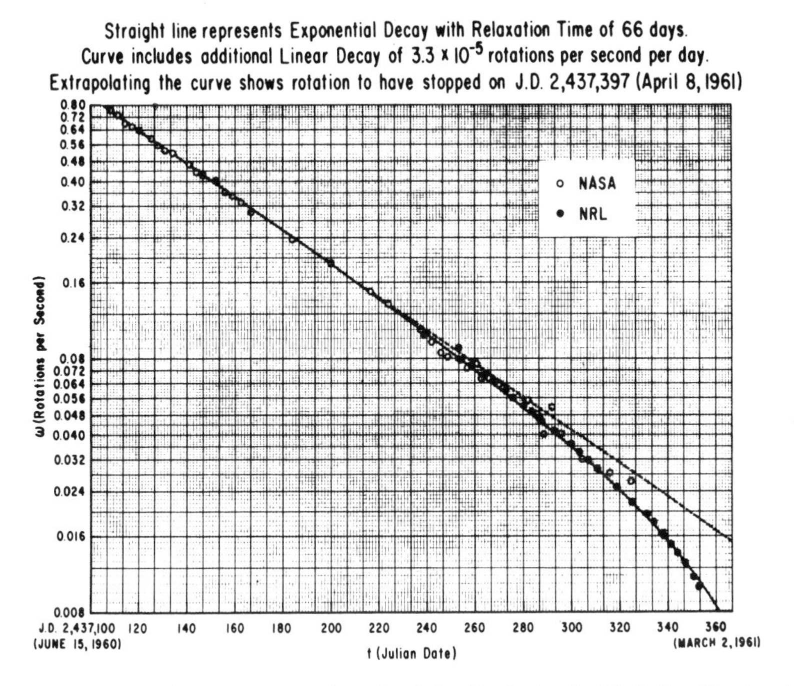

FIG. 9. Observed spin-rate vs. time for Solar Radiation I (1960 Eta 2), showing hysteresis decay as curvature downward from exponential decay line.

Just as, for nonspace ships, reaction with ambient fluids, water or air, is the most practical approach to controlling their rotation, reaction torques on the magnetic fields which fill the ocean of space would always be available, with no expenditure of mass and relatively little of energy, for steering any spaceship with properly designed magnetic rudders.

REFERENCES

1. C. Babbage and J. Herschel, Account of repetition of M. Arago's experiments on magnetism in various substances during rotation. *Phil. Trans. Roy. Soc. London* **115**, Pt. II, pp. 467-496, (1825).

2. H. Hertz, Induction in rotating spheres. *In* "Miscellaneous Papers" (Jones and Schott, transl.) p. 155. Macmillan, London (1896).
3. R. Gans, Über Induktionen in rotierenden Leitern. *Z. Math. Phys.* **48**(1), 1 (1903).
4. E. Stuhlinger, Magnetic Damping of a Spinning Satellite, *Army Ballistic Missile Agency Tech. Memo. RPO-D-TM-3*, August 31, 1956.
5. J. W. Hooper, The Damping of Metallic Cylindrical and Spherical Bodies Rotating in a Uniform Magnetic Field, *Army Ballistic Missile Agency Tech. Rept. DG-R-15*, 1957.
6. K. Jürgensen, N. Sanduleak, and D. Teuber, Experiments in Damping a Rotating Satellite in Magnetic Fields, *ABMA Report DG-M-3-58*, April 22, 1958.
7. J. Vinti, Theory of the Spin of a Conducting Satellite, *Ballistic Res. Labs. Repts Nos. 1020, 1031*, 1957.
8. R. J. Naumann, Recent information gained from satellite orientation measurement. *4th Symp. Ballistic Missiles and Space Technol., UCLA, 1959.*
9. R. H. Wilson, Jr., Magnetic damping of satellite 1958 Beta 2. *Science* **130**, 791 (1959).
10. R. H. Wilson, Jr., Magnetic damping of rotation of the Vanguard I satellite. *Science* **131**, 355 (1960).
11. R. H. Wilson, Jr., Geomagnetic rotational retardation of satellite 1959 Alpha (Vanguard II). *Science* **131**, 223 (1960).
12. R. H. Wilson, Jr., Rotational magnetodynamics and steering of space vehicles. *Proc. XIth Intern. Astronaut. Congr. Stockholm*, Paper 65 (Springer-Verlag, Wien, 1961); also *NASA TN D-566*, 1961.
13. L. H. Grasshoff, A method for controlling the attitude of a spin-stabilized satellite. *Astronautics* Special Section (June 1960).
14. R. E. Fischell, Magnetic damping of the angular motions of earth satellites. *ARS. J.* **31**, 1210 (1961).
15. R. H. Wilson, Jr., Rotational decay of satellite 1960 Eta 2 due to the magnetic field of the earth. *Proc. XIIth Intern. Astronaut. Congr., Washington*, pp. 368-379 (Academic Press, New York, 1963); also *NASA TN D-1469*, 1962.
16. R. H. Wilson, Jr., Optical and Electronic Tracking, *Am. Geophys. Union Monograph No. 4*, p. 67, 1959.
17. V. N. Grigorevskii, Variation of the Period of Rotation of Sputnik 2, *Roy. Aircraft Establishment (Farnborough) Library Translation No. 956*, July 1961.

Magnetic Attitude Control of the Tiros Satellites

E. HECHT and W. P. MANGER

Astro-Electronics Division, Radio Corporation of America, Princeton, New Jersey

I. INTRODUCTION

THE TIROS SERIES of meteorological satellites has been designed to observe and record physical quantities integral to the analysis of earth weather phenomena. Infrared emission, cloud patterns, and cloud distributions over the surface of the earth are monitored by the system. The first view of large-scale weather patterns was provided by the Tiros I satellite.[1] Launched into an almost circular orbit on April 1, 1960, it has since transmitted over 22,000 TV pictures. The satellite was spin-stabilized at 10 rpm so that its axis of rotation would remain in a fixed direction in space. Although it carried three pairs of spin-up rockets, no attitude correction capability was incorporated into the system. All sources of external torque had been considered and it was decided that, except for negligible perturbations, the spin axis would be space stabilized throughout the anticipated three-month operational lifetime.

Two independent television camera systems were mounted in the base parallel to the spin vector but oppositely directed, as shown in Fig. 1. As the satellite progressed in its orbit, the cameras faced either toward or away from the earth. In order to command the vehicle to take pictures only while the cameras were facing earthwards it was necessary to know

[1] The Tiros I Satellite was developed and constructed by RCA under Signal Corps Contract No. DA-36-039-SC-78902. This project was under the management of the National Aeronautics and Space Administration (NASA) and the technical direction of the U.S. Army Signal Research and Development Laboratory. Tiros II and III were produced under NASA Contract Nos. NAS 5-478 and NAS 5-936, respectively.

both its spatial coordinates and its spin-axis attitude. The orientation of the axis of rotation was therefore quite important to the over-all operation of the system. An analysis of the Tiros I vidicon transmissions, showing identifiable landmarks and horizons, indicated that the spin axis, rather than being fixed in space, was moving at a considerable rate. Using the available information at the time, an immediate investigation was undertaken (by personnel of NASA Goddard Space Flight Center and RCA) to determine the causes of this motion.

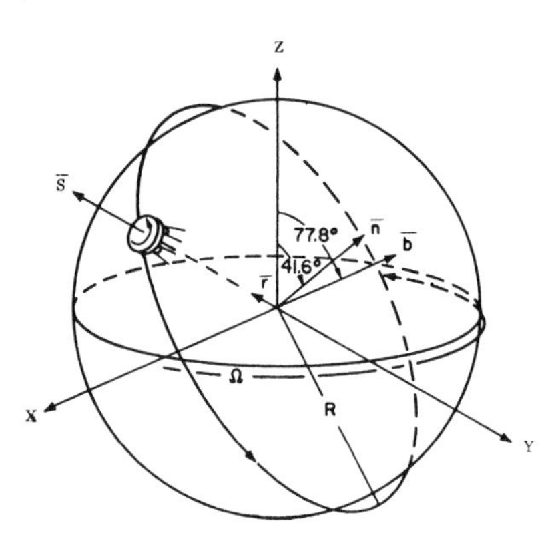

FIG. 1. Tiros I at injection, 1152 GMT, April 1, 1960, X, Y, and Z are space coordinates with origin at the center of the earth and with the X-Y plane in the equatorial plane of the earth. **s** is the unit spin vector of the satellite. The astronomical coordinates of **s** are declination, + 19.8 deg; right, ascension, + 58.6 deg. Ω is the angle from the X axis to the orbital ascending node. The orbital nodes regress at the rate of -4.547 deg day^{-1} (westward). **n** is the unit vector normal to the orbit; **b** is the mean normalized magnetic dipole field vector appearing to the satellite in one orbit; **r** is the unit vector from the center of the earth toward the satellite. R is the orbital radius.

II. PERTURBING TORQUES

The characteristics of the major sources of external torque which might affect a satellite of the configuration of Tiros I (a short cylinder, 22.5 inches high and 42 inches in diameter) are both gravitational and magnetic. A nonsymmetric, rigid body in a gravitational field will experience a measurable torque if its principal moments of inertia are unequal. The moment of inertia about the satellite spin axis was larger than the other principal moments of inertia. The magnitude of the resulting differential gravity torque was investigated and, although it influenced the satellite, it could not account for the observed motion. It then became apparent that the primary perturbation was magnetic in origin, namely, a satellite dipole interaction with the earth's magnetic field. Such a dipole might be the result of closed current loops in the instrumentation or permanent magnetism of ferromagnetic materials in the vehicle. Using a dipole model for the earth's field and including the secondary effects of differential gravity, the mathematics describing the motion were developed for computer programming. The equations employed had their basis in angular momentum considerations, since the applied torque equals the rate of change of angular momentum. A number of simplifying approximations were utilized resulting in a considerable reduction of computer running time. The equations used had the following form:

$$\frac{d\mathbf{s}}{dt} = \tfrac{1}{2}\epsilon(\mathbf{s}\cdot\mathbf{n})(\mathbf{s}\times\mathbf{n}) + \mu(\mathbf{b}\times\mathbf{s})$$

where

$$\epsilon = \frac{3\omega_0^2}{\omega_s}\frac{(I-J)}{I}$$

and

$$\mu = \frac{M}{I\omega_s}\frac{(V_0)}{R^3}$$

$\mathbf{s}$ = Unit spin vector

ω_0 = Angular velocity of the orbital radius vector

ω_s = Angular velocity of the satellite about the spin axis

I = Moment of inertia

J = Transverse moments of inertia

$\mathbf{n}$ = Unit vector normal to the orbit

M = Magnetic dipole moment along the spin axis

V_0 = Magnetic constant for a dipole at the center of the earth

R = Orbital radius

$\mathbf{b}$ = Normalized magnetic dipole field vector for the earth's field

In order to reflect the observed decrement in spin rate from 10.0 rpm on April 1 to 9.4 rpm on May 27, an additional eddy-current term was introduced. The integration of these equations by RCA 501 digital computer resulted in very fine agreement with the observed motion. Figure 2 is a plot of the experimental data obtained from photogrammetric attitude analysis, and the corresponding computer values. The predictive capabilities of the program indicated the feasibility of a magnetic attitude control system which would direct the course of the spin axis through the earth's magnetic field.

Shortly after the successful launch of the Tiros I satellite, RCA was awarded a contract by NASA to modify the duplicate test and back-up models so that they could accommodate a NASA-developed infrared experiment. Along with this new equipment Tiros II contained a

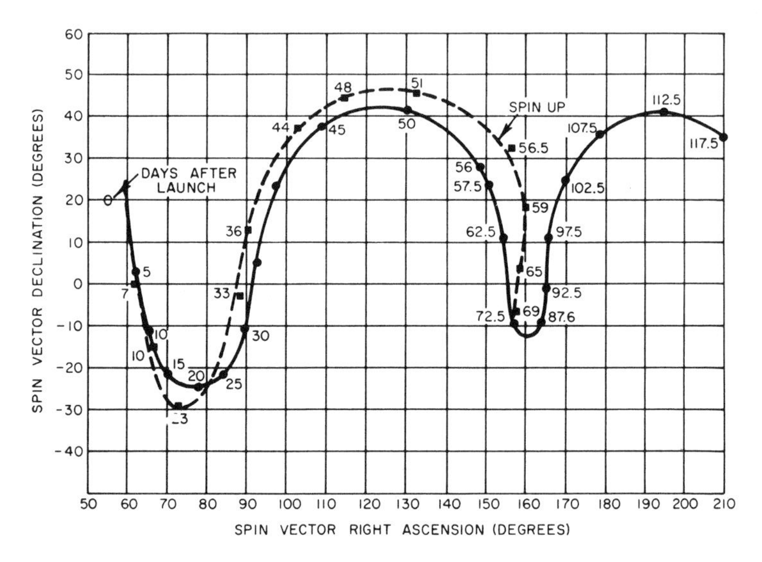

FIG. 2. Observed motion of the Tiros I spin vector based on an analysis of photographs (dashed line) compared with the theoretical motion based on the effects of a magnetic dipole moment along the spin axis and differential gravity (solid line). Declination is + north and − south of the celestial equator. Right ascension is + east of the vernal equinox along the celestial equator. Increased stability of both the theoretical and observed motion is seen after spin-up on May 27 (day 56 after launch). The last picture with clearly identifiable landmarks was received on June 9 (day 69 after launch).

magnetic attitude control system designed to modify the satellite's spatial orientation so as to obtain optimum performance from the infrared subsystem, cameras, and solar power supply. The developmental procedure applied to the attitude control system followed two major paths: (1) it was deemed desirable to reduce the residual dipole moment as much as possible so as to mitigate the effects of uncontrollable torques; and (2) the means by which known dipole moments could be programmed into the satellite were evaluated. The reduction of the Tiros II dipole moment required a knowledge of both its magnitude and direction in each of the satellite's operational modes. This information was obtained through the use of the test apparatus shown in Fig. 3. The satellite was spun in a space surrounded by two spherically wound coils. One of the coils was used to cancel the earth's magnetic field. Currents induced in the other coil were proportional to the spinning dipole within the satellite. The order of magnitude of the measured Tiros II dipole moment compared quite favorably with the value computed for Tiros I. A small permanent magnet was then incorporated in the satellite to effectively cancel its residual magnetic dipole.

FIG. 3. Satellite dipole-moment measuring facility.

The active control element was a coil of wire wound about the periphery of the satellite. The magnitude and direction of currents through the coil provided both positive and negative magnetic moment compensation at various levels ranging from $+3$ to -3 amp-turns-meter^2. The desired currents were introduced via a stepping switch which was activated by an auxiliary control system.

III. Magnetic Attitude Control Electronics

The electronics for the system served to provide both a means of changing the impressed currents in the control coil and a telemetry readout of the position of the Magnetic Attitude Control (MAC) switch. Initial studies of the over-all attitude correction function indicated that

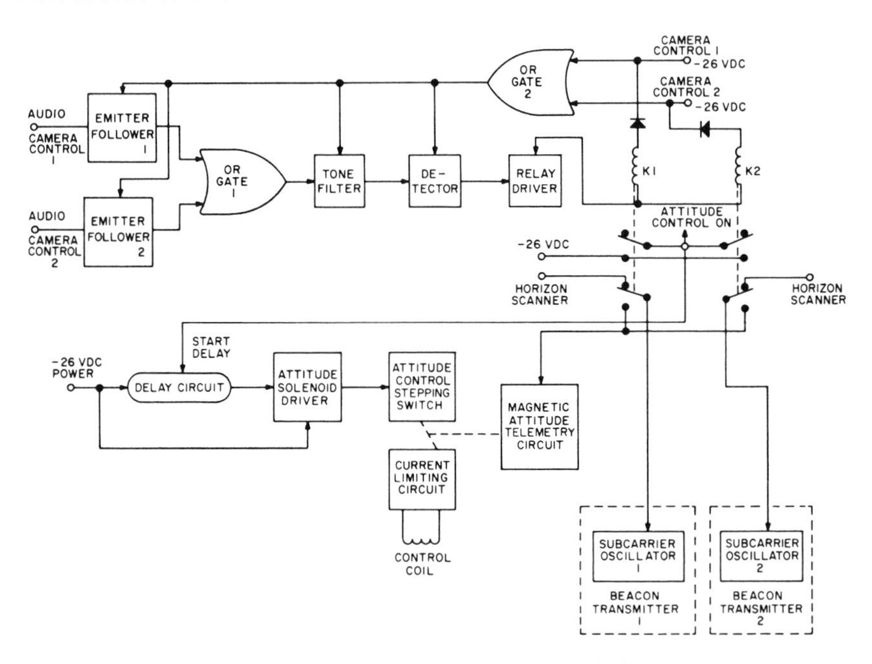

Fig. 4. Attitude control electronics, block diagram.

programming would not be required at frequent intervals. A single pulse would, therefore, provide adequately reliable control over the stepping switch.

An operating voltage was applied to the attitude correction electronics from the control relays in one of the two camera systems being programmed by the ground station (see Fig. 4). This took place approximately 28 seconds after the start of a direct camera or playback sequence. The control tones were passed through emitter followers 1 and 2 which provided isolation between the two camera control packages while also allowing for redundancy. After passing through the OR gate 1, the signal was applied to a filter which selected the attitude command frequency and fed it into a pulse detector. The detector turned on a relay driver which in turn energized either relay K1 or K2 depending upon which camera system was in use. With the activation of either of the relays the magnetic attitude telemetry circuit was connected to a beacon transmitter by way of the subcarrier oscillator. The attitude telemetry circuit had a dc voltage output which was a function of the position of the MAC switch. This monitoring of the switch position was done in less than 6 seconds. When the command tone was received for more than 6 consecutive seconds the delay circuit was triggered and the MAC switch was advanced one position clockwise thereby changing the current in the attitude coil. In this manner one command tone controlled both the monitoring and changing of the satellite's dipole moment.

IV. MAC System Experimental Results

At 6:13 A. M. on November 23, 1960, Tiros II was placed into orbit with an apogee of 453 statute miles. The satellite functioned for 12 months, far exceeding its designed lifetime of 90 days, during which time 35,000 pictures were received from its two cameras. The magnetic attitude control system was effectively applied throughout this period. Unfortunately, difficulties in data acquisition and reduction as well as a number of randomly unprogrammed changes in the MAC switch position limited the full capability of the system.

Three considerations required close moderation of the Tiros III spin axis attitude in space. Because of power demands, the orientation of the satellite's 9300 solar cells had to be maintained within set limits. The infrared sensors could not be allowed to point toward the sun for any extended period. Finally, accurate analysis of cloud-cover photographs necessitated the inclusion of earth horizons in the pictures.

The design of Tiros III was augmented to remove the difficulties incurred in Tiros II. Since its launch on July 12, 1961, the MAC switch has been changed nineteen times, resulting in excellent control of the spin-axis orientation and completely satisfying all of the above requirements. Figures 5 and 6 are graphs of the right ascension and declination of the spin axis normal point on a per orbit basis. The points are the observed values and the crosses are the computer predictions. Similar

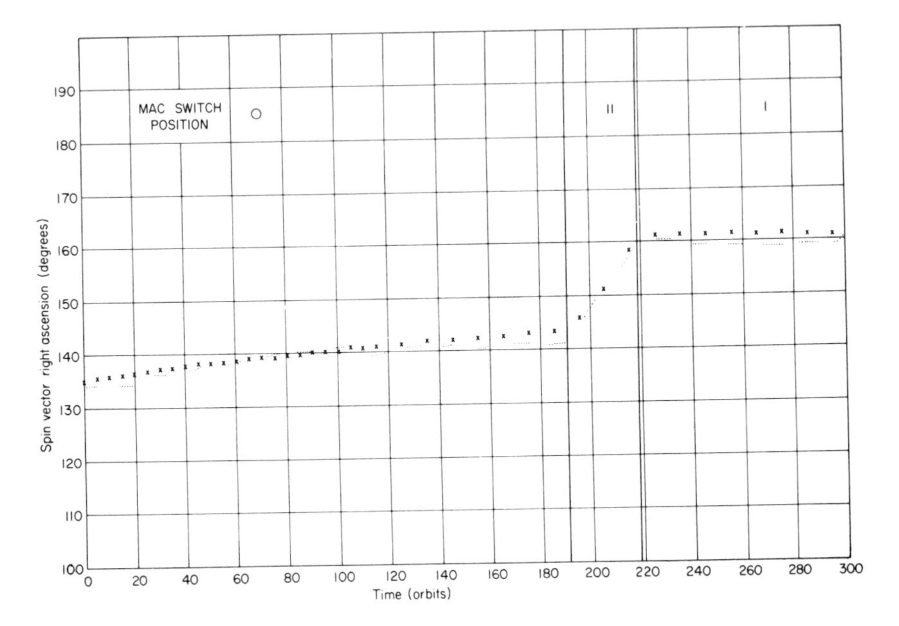

FIG. 5. Orbits 0-300 normal point right ascension.

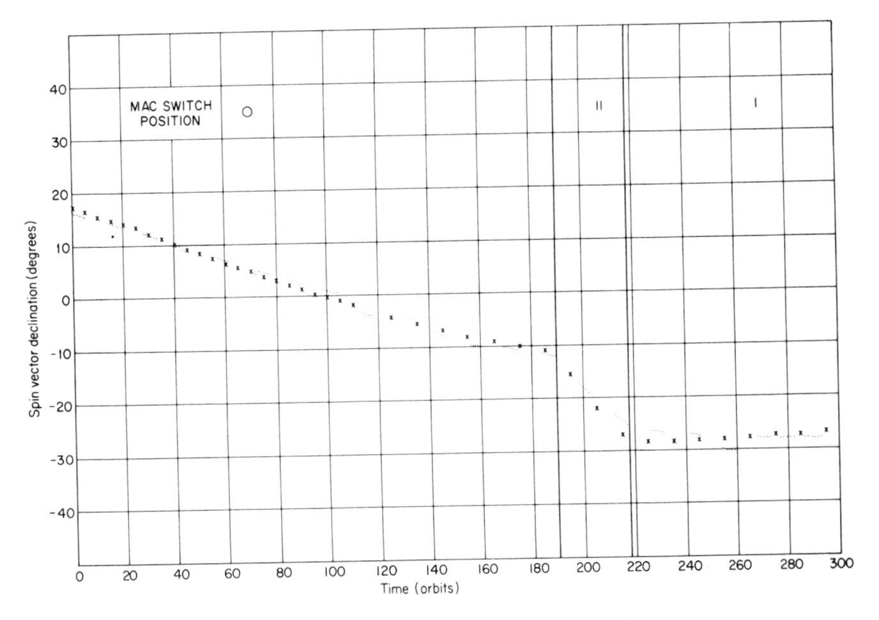

FIG. 6. Orbits 0-300 normal point declination.

capabilities have been developed at NASA's Goddard Space Flight Center and at RCA Astro-Electronics Division where the satellite's entire attitude history (over 2000 orbits, to date) has been prognosticated with great accuracy. Spin rates, sun angles, and all of the other desired quantities have been calculated with similar facility.

The magnetic torquing system has reached a degree of perfection which clearly warrants its inclusion in all future Tiros satellites. Similar systems will surely find useful application in other satellites plying the earth's magnetic field.

References

1. W. Bandeen and W. P. Manger, "Angular Motion of the Spin Axis of the TIROS I Meteorological Satellite Due to Magnetic and Gravitational Torques," *J. Geophys. Res.*, September 1960.
2. R. A. Nidey, Gravitational torque on a satellite of arbitrary shape. *ARS J.* February 1960.
3. O. E. Rosner and W. P. Manger, Spin Axis Motion of TIROS I Meteorological Satellite Caused by Magnetic, Gravitational, and Eddy Current Torques, *TIROS I Post-Launch Evaluation Rept.*, July 1, 1961.
4. S. Dreskin, Mathematical Analysis of the Spin Axis Motion of TIROS II Meteorological Satellite, *RCA-AED Internal Rept.*

Satellite Angular Momentum Removal Utilizing the Earth's Magnetic Field

ROBERT J. MCELVAIN[1]

Space-General Corporation,
El Monte, California

I. Introduction

MANY DISTURBANCE TORQUES acting on satellite vehicles are due to interactions of the vehicle with various ambient fields. In some cases, it is feasible to control the interactions of the vehicle with one or more of these ambient fields so that the disturbance torque becomes a control torque for the vehicle. One application of this method is to generate a controlled magnetic moment in a satellite vehicle which interacts with the earth's magnetic field in order to control the vehicle orientation, or to perform some similar control function.

The advantage of utilization of controlled interactions with ambient fields for vehicle control purposes is that no fuel need be carried aboard the vehicle as is necessary for a mass expulsion system. The only requirement for the generated magnetic moment is electrical power, which can be obtained from solar cells; hence, the vehicle lifetime is not limited by the fuel storage capability. However, in some cases, the low-level torques obtainable from vehicle interactions with ambient fields may not be sufficient to perform the required control maneuvers. For angular momentum removal from a momentum storage device, such as reaction wheels, it will be seen that the obtainable magnetic

[1] Present address: Hughes Aircraft Co., Space Systems Division, El Segundo, California.

torque capability will in general be quite sufficient for low-altitude earth orbits. This paper will be concerned with the formulation of the magnetic control scheme for momentum dumping only, assuming that the vehicle will contain such a momentum storage device.

II. Coordinate Systems and Notation

Definition of several reference coordinate sets is convenient in order to express the orientation of the vehicle relative to inertial space, and to express the magnetic field as seen by the satellite. These coordinate sets are shown in Figs. 1 and 2, and are defined as follows:

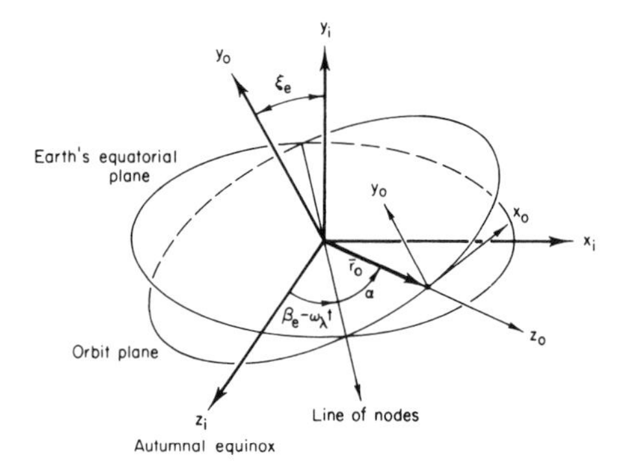

Fig. 1. Orbital and inertial reference coordinate sets.

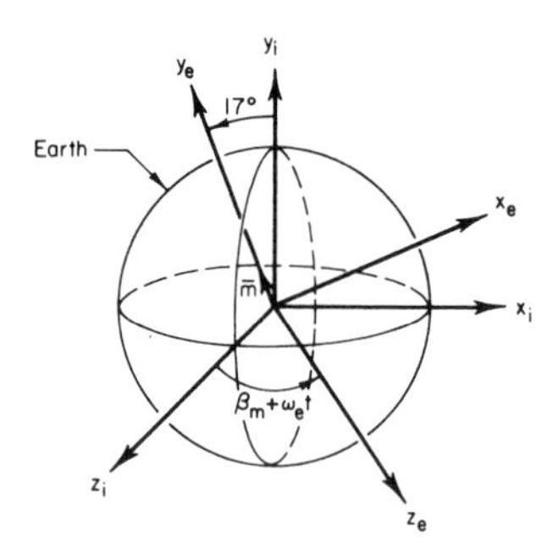

Fig. 2. Magnetic reference coordinate system.

1. Inertial reference $(x_i y_i z_i)$: An inertial coordinate set with origin at the center of the earth; the $x_i z_i$ plane the earth's equatorial plane; and the z_i axis along the line of equinoxes, directed toward the autumnal equinox.

2. Orbital reference ($x_0y_0z_0$): A coordinate set with origin at the center of the earth; the x_0z_0 plane the orbit plane (which in general regresses in inertial space); and the z_0 axis along the radius vector, $\mathbf{r}_0$, directed from the center of the earth to the vehicle center of mass (c.m.). The angle ξ_e is the inclination of the orbit plane to the equatorial plane. The angle between the line of nodes (intersection of the orbit plane with the equatorial plane) and the z_i axis is $\beta_e - \omega_\lambda t$, where β_e is this angle at $t = 0$, and ω_λ is the nodal regression rate of the orbit plane. The angle α defines the angular position of the satellite relative to the ascending node.

3. Geomagnetic reference ($x_ey_ez_e$): An earth-fixed reference coordinate set with origin at the center of the earth; the y_e axis directed toward the magnetic Noth Pole; the x_ez_e plane the geomagnetic equatorial plane; and the z_e axis along the intersection of the geomagnetic equatorial plane and the earth's equatorial plane. The angle between the z_e axis and the z_i axis is $\beta_m + \omega_e t$, where β_m is this angle at $t = 0$ and ω_e is the earth's rotation rate.

In addition, a body-fixed coordinate set xyz is defined with origin at the vehicle center of mass. The xyz coordinate may be chosen to be the principal axes of the vehicle if convenient. For an attitude-controlled satellite, the desired orientation of the vehicle relative to a convenient reference is known as a function of time. Since many applications of earth satellites require alignment of a body-fixed reference with the radius vector, $\mathbf{r}_0$, a convenient choice for a reference is the $x_0y_0z_0$ coordinate set. The orientation of the body set relative to the $x_0y_0z_0$ set may be uniquely defined by three angles, ψ, θ, and ϕ. These angles are represented by ordered rotations about the z_0, the y_0', and the x_0'' axes, where the primes refer to the new orientation of the coordinate axes after rotations. The matrix [A] which resolves vectors in the $x_0y_0z_0$ set to the body set is given by:

$$[A] = \begin{bmatrix} a_{11} & a_{12} & a_{13} \\ a_{21} & a_{22} & a_{23} \\ a_{31} & a_{32} & a_{33} \end{bmatrix} \tag{1}$$

where

$$\begin{aligned}
a_{11} &= \cos\theta\cos\psi \\
a_{12} &= \cos\theta\sin\psi \\
a_{13} &= -\sin\theta \\
a_{21} &= -\sin\psi\cos\phi + \sin\phi\sin\theta\cos\psi \\
a_{22} &= \cos\phi\cos\psi + \sin\phi\sin\theta\sin\psi \\
a_{23} &= \sin\phi\cos\theta \\
a_{31} &= \sin\psi\sin\phi + \sin\theta\cos\psi\cos\phi \\
a_{32} &= -\cos\psi\sin\phi + \cos\phi\sin\theta\sin\psi \\
a_{33} &= \cos\theta\cos\phi
\end{aligned} \tag{2}$$

The angles ϕ, θ, and ψ can be written as functions of time or orbital position α to define the prescribed vehicle orientation relative to $x_0y_0z_0$. With the aid of these definitions, a mathematical model of the earth's magnetic field is developed in Appendix I. The development is based on the assumption that the earth's magnetic field can be sufficiently represented as that due to a magnetic dipole with dipole axis inclined 17 deg to the y_i axis.

III. Basic Magnetic Control Concepts

The interaction of the earth's magnetic field with magnetic moments fixed or generated within the satellite vehicle will produce torques on the vehicle. This torque is given by:

$$\mathbf{T} = \mathbf{M} \times \mathbf{B}, \tag{3}$$

where $\mathbf{M}$ is the generated magnetic moment, and $\mathbf{B}$ is the earth's magnetic field intensity. The torque developed can then be used for control purposes, provided that a suitable control law for $\mathbf{M}$ can be chosen. Let $\mathbf{H}_w$ represent the unwanted portion of the momentum of the momentum storage device; then, if a proportional system is considered, the magnetic torque required to remove this portion is given by:

$$\mathbf{T} = -K\mathbf{H}_w \tag{4}$$

where K is an arbitrary constant. Equating expressions (3) and (4) and taking the vector cross product of $\mathbf{B}$ with both sides results in:

$$\mathbf{B} \times (-K\mathbf{H}_w) = \mathbf{B} \times (\mathbf{M} \times \mathbf{B}) = B^2\mathbf{M} - \mathbf{B}(\mathbf{M} \cdot \mathbf{B}) \tag{5}$$

From Eq. (3), the maximum torque for a given magnitude of $\mathbf{B}$ and $\mathbf{M}$ is obtained when $\mathbf{M}$ is normal to $\mathbf{B}$; hence, from Eq. (5), a control law for $\mathbf{M}$ is implied of the form:

$$\mathbf{M} = -\frac{K}{B^2}(\mathbf{B} \times \mathbf{H}_w) \tag{6}$$

This moment then produces a torque, $\mathbf{T}$, given by:

$$\mathbf{T} = -\frac{K}{B^2}[B^2\mathbf{H}_w - \mathbf{B}(\mathbf{B} \cdot \mathbf{H}_w)] \tag{7}$$

Physically, Eqs. (3), (6), and (7) state that no torque can be obtained about the $\mathbf{B}$ vector. Hence, when $\mathbf{H}_w$ is not normal to $\mathbf{B}$, the control law will produce a torque which is the projection of $-K\mathbf{H}_w$ in the plane normal to $\mathbf{B}$. This torque will then have a component along the desired direction and an error component normal to this direction; therefore, it is necessary to determine the convergence of three-axis magnetic momentum dumping for a particular satellite mission. In this concept, convergence is defined as the capability of the magnetic control scheme to remove excess angular momentum from the vehicle such that a desired vehicle angular momentum can be maintained in order to fulfill the vehicle orientation requirements.

In general, both the direction and magnitude of $\mathbf{B}$, as seen in body-fixed axes, will vary during an orbit. Then, the average value of the term $(\mathbf{B} \cdot \mathbf{H}_w)$ will be less than or equal to the product of the magnitudes, BH_w. The unique case, where $\mathbf{H}_w$ varies so as to always be parallel to $\mathbf{B}$, will produce no torque, hence, no angular momentum can be removed or added to the vehicle due to magnetic torquing. This unique case is physically quite unrealistic owing to the usual inability of a magnetic torquing system to provide large enough torques to change the direction of the total vehicle angular momentum as rapidly as $\mathbf{B}$ changes. Therefore, $(\mathbf{B} \cdot \mathbf{H}_w)$ will in general be less than BH_w, and the torque will always act such as to reduce the magnitude of $\mathbf{H}_w$. The convergence of magnetic momentum removal is shown in the vector diagram of Fig. 3, for a finite correction time, Δt, when $\mathbf{B} \cdot \mathbf{H}_w \neq BH_w$.

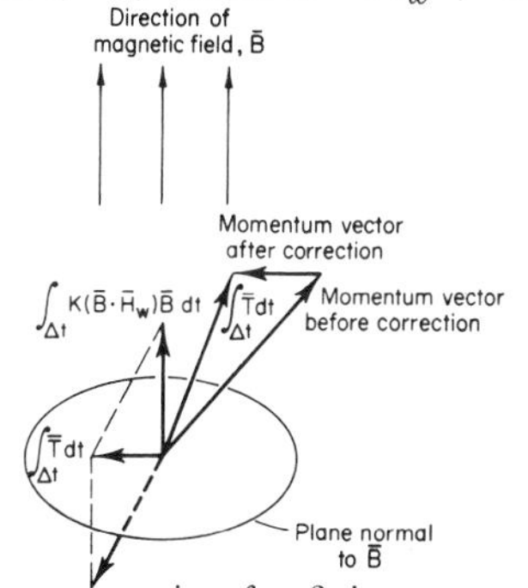

FIG. 3. Vector representation for finite momentum correction.

The magnetic moment required by Eq. (6) may be generated on board the vehicle by use of current coils or electromagnets rigidly attached to the vehicle. The components of moment required by the control law [Eq. (6)], expressed in body coordinates are:

$$\begin{bmatrix} M_x \\ M_y \\ M_z \end{bmatrix} = -\frac{K}{B^2} \begin{bmatrix} B_y H_{wz} - B_z H_{wy} \\ B_z H_{wx} - B_x H_{wz} \\ B_x H_{wy} - B_y H_{wx} \end{bmatrix} \tag{8}$$

The magnetic moment, **M**, due to a planar current loop with current i, N turns, and area A, is given by:

$$\mathbf{M} = iNA\hat{n} \tag{9}$$

where $\hat{n}$ is a unit vector directed normal to the area A. Similarly, a magnetic moment may be produced by current coils with a bar magnet core, which gives an effective gain in the magnitude of the moment relative to a coil with air core of the same area, number of turns, and current. Then if three orthogonal body-fixed current coils (or electromagnets) with normals along the body x, y, and z axes are assumed, the required currents in the coils are given by:

$$i_x = -\frac{K}{B^2 N_x A_x}(B_y H_{wz} - B_z H_{wy}) \tag{10}$$

$$i_y = -\frac{K}{B^2 N_y A_y}(B_z H_{wx} - B_x H_{wz}) \tag{11}$$

$$i_z = -\frac{K}{B^2 N_z A_z}(B_x H_{wy} - B_y H_{wx}) \tag{12}$$

where $N_i A_i (i = x, y, z)$ are the number of turns and coil areas along the respective axes.

From Eqs. (10) through (12), it can be seen that the currents required depend on the magnitude (squared) of the earth's **B** field, B^2. In most satellite orbits, B^2 is a time-varying function; however, to avoid computing this quantity during corrections, a constant value of B^2 may be used in the control law. This simplification appears as a time-varying gain of the momentum-removal system as seen from Eq. (6); hence, the effective response time of the system will vary at different orbital positions. Then, if the constant value $B_0{}^2$ is chosen such that $B_0{}^2$ is the minimum value of $B^2(t)$ experienced during the vehicle lifetime, the system response time will always be equal to or less than that for $B_0{}^2$. The system constants (number of turns, current, etc.) can then be sized in order to achieve a desired minimum response characteristic. For a circular orbit, the maximum variation of $B^2(t)$ is a factor of 4; hence, the actual system response could be as much as four times faster than the minimum case. Similar response time limits can be determined for an elliptical orbit.

The actual implementation of the basic control law for a particular satellite vehicle can result in a variety of schemes. The differences between the various schemes will be due to constraints imposed by the particular satellite mission. Several typical constraints would be the following:

1. Magnetic torquing should occur only when the term $(\mathbf{B} \cdot \mathbf{H}_w)$ is a minimum, in order to reduce the momentum errors coupled into the axes transverse to the desired removal direction.

2. The magnetic torquing system should not operate continuously in order to avoid interference of the generated magnetic field with experiments.

3. The signal, $\mathbf{H}_w$, from the momentum storage device, should operate in a threshold manner such as to prevent unnecessary removal of angular momentum which is periodic in inertial space.

4. Weight, power requirements, and reliability of the system should be optimized, consistent with the over-all vehicle mission.

Other similar constraints can be envisioned which would influence the design of the magnetic torquing system implementation. Based upon these typical constraints, three distinct implementations for magnetic momentum removal will evolve. These schemes may be summarized as follows:

a. *The Continuous Scheme.* Corrections are made continuously, and the moment $\mathbf{M}$ is varied continuously as the $\mathbf{B}$ field changes and as the momentum storage device momentum, $\mathbf{H}_w$, changes.

b. *The Intermittent Scheme.* Corrections are made intermittently and occur only when $\mathbf{H}_w$ exceeds a fixed preset value. During any one correction, the moment $\mathbf{M}$ is varied continuously as the $\mathbf{B}$ field varies. The momentum increment removed will be constant for successive corrections, but the duration of successive corrections will vary.

c. *The Optimum Intermittent Scheme.* Corrections are made intermittently and occur only when $\mathbf{H}_w$ has exceeded a fixed preset value, and the term $(\mathbf{B} \cdot \mathbf{H}_w)$ is minimum during the orbit. When both conditions occur, the components of $\mathbf{B}$ are sampled and held,[2] and a constant moment $\mathbf{M}$ is generated for a fixed length of time. The momentum increment removed will vary between successive corrections, but the duration of each correction is the same.

IV. General Control System, Description and Implementation

A block diagram showing the utilization of the general magnetic momentum removal system in the over-all control system design is

[2] The sampling is not mandatory, as the moment can be varied as $\mathbf{B}$ varies. However, in general, the correction duration is short so that $\mathbf{B}$ will not vary appreciably during any one correction.

shown in Fig. 4. The momentum storage device is assumed, for convenience, to be three orthogonal reaction wheels, although the discussion of momentum removal is independent of the type of momentum storage device. The wheels may be driven by any type of control law, such as proportional, proportional plus integral, etc.; and the wheels may operate in either a continuous or bang-bang manner.

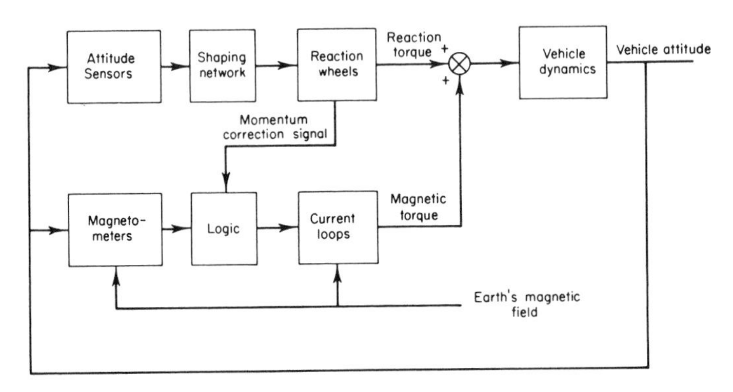

FIG. 4. Block diagram representation of typical vehicle control system.

Attitude error information may be derived from a variety of sensors, depending on the vehicle orientation requirements. Examples of possible sensors are rate and position gyros, sun sensors, horizon scanners, star trackers, etc.

The logic block shown in Fig. 4 for the magnetic momentum removal loop represents the actual implementation of the control law and associated logic for the particular scheme used. This block will be discussed in some detail for each of the three basic schemes: the Continuous scheme; the Optimum Intermittent scheme, and the Intermittent scheme.

The detailed implementation of the Continuous scheme is shown by the block diagram of Fig. 5. The outputs of the magnetometer along the three axes are multiplied and summed with the three reaction wheel momentums according to Eq. (6). These resulting signals represent the current commands to the three current coils (or electromagnets) as functions of time. The gains shown in the block diagram represent the physical size of the coils; that is, the number of turns with a given area that will produce a specified torque in a magnetic field B_0^2, for a specified value of wheel momentum. In actuality, the gains will also consist of such hardware calibrations as the number of volts per gauss, the number of volts per foot-pound-second, the coil current per volt, etc.

Although the basic implementation appears relatively simple, the Continuous scheme has several disadvantages:

1. Continuous operation may produce low reliability of system operation for extended lifetimes.

2. The circuitry which performs the multiplication and summation must be built to operate on variable signals from both the magnetometers and the wheels. This implies the use of electronic multipliers, as opposed to summing amplifiers in the other schemes. In general, the latter will be more reliable.

3. The momentum removal efficiency is less than that for the Optimum Intermittent scheme.

4. The magnetometer must be shielded so that the continuously generated magnetic field does not interfere with the measurement of the earth's **B** field.

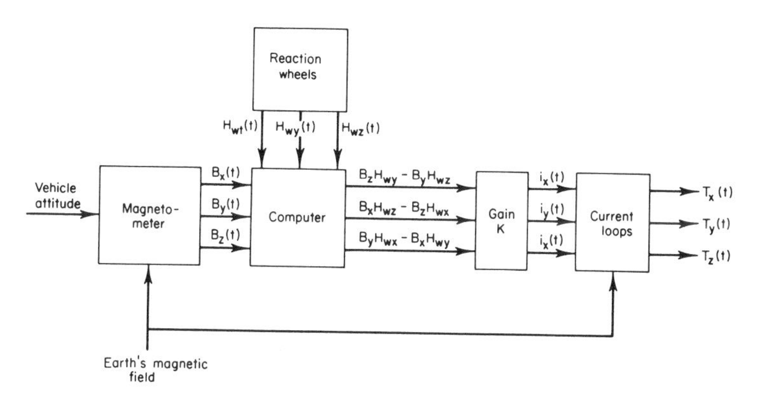

FIG. 5. Block diagram representation of continuous momentum dumping scheme.

The block diagram shown in Fig. 6 represents the implementation of the Optimum Intermittent scheme. This scheme is the most efficient of the three schemes for magnetic momentum removal. However, the one serious drawback to the implementation of this scheme lies in the difficulty of logic determination of the occurrence of the condition $(\mathbf{B} \cdot \mathbf{H}_w)$ minimum for all types of orbits and vehicle orientation requirements. This condition obviously occurs when the component of **B** is zero along the axis in which the correction is to be made. For this case, simple switches can be placed in series with the magnetometer so that the output is zero unless the **B** field component is zero or near zero. It is this method that is represented in Fig. 6. Alternate methods to determine the minimum condition when each **B** field component minimum is nonzero would usually require a rather complex memory and logic device.

The sampler shown in Fig. 6 samples the **B** field components when a correction is called for and holds the value for a fixed length of time Δt. In some very special orbits, such as an earth-tracking satellite in a circular earth equatorial orbit, the **B** field components can be preprogrammed, hence eliminating the sampler, and requiring the magnetometers only to determine the minimum condition of $(\mathbf{B} \cdot \mathbf{H}_w)$.

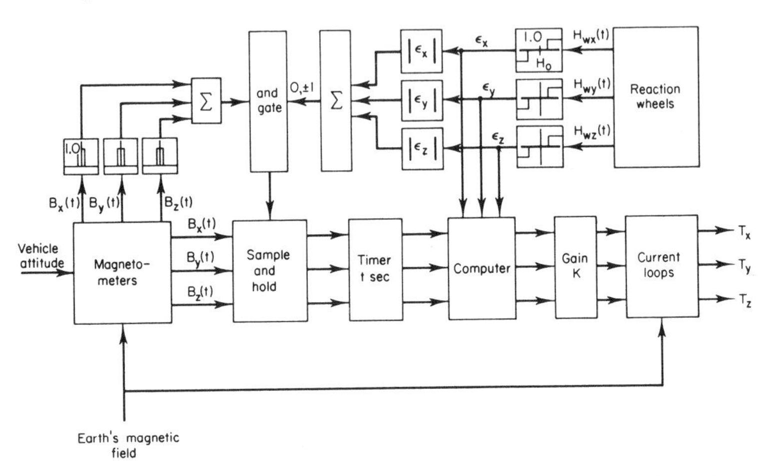

FIG. 6. Block diagram representation of optimum momentum dumping scheme.

The gain shown in the block diagram again represents the physical size of the current loop, such that when $(\mathbf{B} \cdot \mathbf{H}_w) = 0$, a correction of a fixed magnitude will occur in Δt seconds in a field strength of B_0^2.

The implementation of this scheme has several disadvantages as a result of the foregoing discussion:

1. System reliability may be degraded by the logic requirements necessary to determine the minimum condition in the general case.

2. Satellite orientation requirements which do not entail an inertially fixed orientation cause delays in corrections owing to the fact that an inertial angular momentum vector appears as a rotating vector as seen by the body-fixed wheels. Therefore, the corrections can occur only when the minimum condition is satisfied at the same time that the momentum vector appears in the proper position for removal. If both the **B** field components and the momentum vector vary at the same frequency, and are in phase, corrections are impossible.

The major advantages to utilization of this scheme are:

a. Momentum removal is the most efficient that is physically realizable.

b. The corrections are intermittent and of fixed duration; hence estimates of duty cycles may be made. Also, continuous interference of the generated field with experiments on the vehicle is avoided.

c. The circuitry necessary to obtain current commands as in Eqs. (10) through (12) is quite simple, requiring only amplifiers instead of electronic multipliers.

d. Wheel momentum is not driven to zero; hence power will not be wasted removing inertially periodic momentum as in the Continuous scheme.

Several disadvantages of both the Continuous scheme and the Optimum Intermittent scheme can be eliminated by use of the Intermittent scheme. The implementation of the scheme is represented by the block diagram of Fig. 7. As in the Optimum Intermittent scheme, corrections are performed only when the wheel momentum exceeds a preset value;

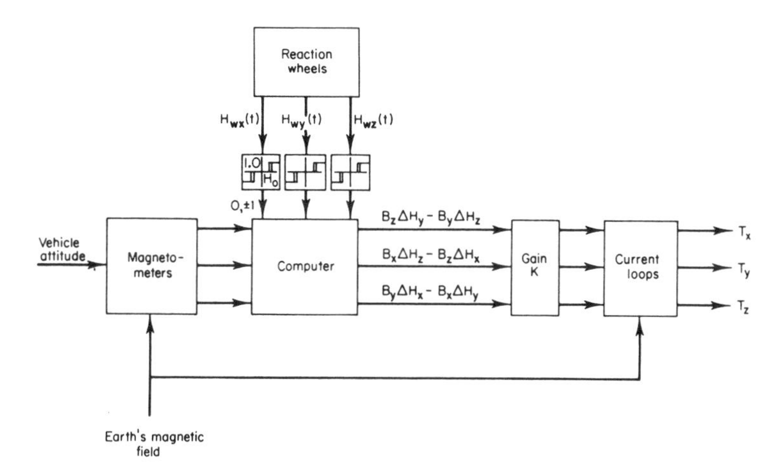

FIG. 7. Block diagram representation of the intermittent momentum dumping scheme.

however, neither the condition that $(\mathbf{B} \cdot \mathbf{H}_w)$ be minimum nor the sampling and holding of the magnetometer outputs is required. The wheel momentum signals appear only in the three discreet values in each axis: 0, $\pm H_0$, where H_0 is the preset value of the wheel momentum in a particular axis. These wheel momentum signals are multiplied and summed according to Eqs. (10) through (12) to provide the current commands to the three coils. The gains shown in the block diagram represent the physical sizing of the loops as in the other two schemes. The Intermittent scheme loops are sized so that, when $(\mathbf{B} \cdot \mathbf{H}_w) = 0$, a momentum correction of ΔH will be completed in Δt seconds. In actuality, the corrections will take more than Δt seconds, as the logic does not determine the occurrence of $(\mathbf{B} \cdot \mathbf{H}_w)$ minimum.

The disadvantages of the Intermittent scheme can be summarized as follows:

1. The efficiency of momentum removal over a large number of corrections is no better than that of the Continuous scheme.

2. Although the corrections take place intermittently, the time to complete any one correction is variable.

3. The magnetometers must be shielded to avoid interference during corrections as in the Continuous scheme.

The advantages to the implementation of this scheme are:

a. The corrections are intermittent; hence possible interference of the generated field with experiments on the vehicle can be avoided.

b. The sometimes complex logic necessary to determine the $(\mathbf{B} \cdot \mathbf{H}_w)$ minimum condition is not necessary. This scheme is especially less complex for orbits in which the components of $\mathbf{B}$ are never zero.

c. Wheel momentum will not be driven to zero as in the Continuous scheme; hence power will not be consumed to remove periodic inertial momentum.

d. The sampler necessary in the Optimum Intermittent scheme is no longer required.

e. The circuitry which performs the multiplication and summation can consist of summing amplifiers instead of electronic multipliers as in the Continuous scheme.

In general, the choice of the best scheme to use will depend upon the particular vehicle requirements, and trade-offs between weight, power, reliability, and efficiency of the various schemes.

Detailed design applications of the Continuous scheme are given in references [1, 2]. As an example of design procedure, the utilization of one of the intermittent schemes for a particular satellite vehicle is treated in this paper.

V. Magnetic Momentum Removal for an Earth-Tracking Satellite

The vehicle chosen to demonstrate the design procedure is an earth-tracking vehicle in a 300-nm circular orbit. An intermittent scheme requirement will be assumed in order to avoid possible interference of the generated magnetic field with experiments. The design procedure will demonstrate the analysis leading to the choice of one of the two intermittent schemes for this particular vehicle.

The earth's magnetic field intensity will vary between 0.243 gauss and 0.486 gauss for a 300-nm circular orbit. The mathematical expression for the components of the dipole field model is given by Eq. (A12) of the Appendix:

$$\begin{bmatrix} B_{x0} \\ B_{y0} \\ B_{z0} \end{bmatrix} = \frac{m}{r_0^3} \begin{bmatrix} \sin \xi_m \cos (\omega_e t - \eta_m) \\ \cos \xi_m \\ -2 \sin \xi_m \sin (\omega_e t - \eta_m) \end{bmatrix} \tag{A12}$$

The motion of this vector as seen in the $x_0 y_0 z_0$ reference set is shown in Fig. 8. The vector is seen to trace out an elliptical cone with axis of symmetry along the y_0 axis.

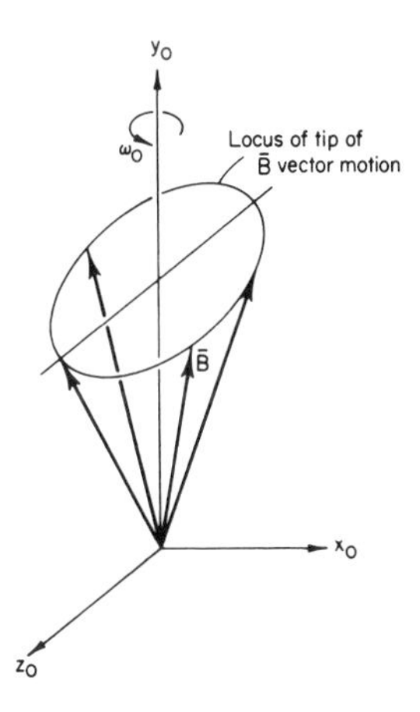

FIG. 8. Representation of earth's magnetic field as viewed in orbital reference coordinate set.

For simplicity, the desired orientation of the vehicle body axes will be assumed to be the $x_0 y_0 z_0$ reference set. The vehicle is assumed to contain a set of three orthogonal reaction wheels for momentum storage and precise attitude control. Attitude signals are assumed to exist which will drive the reaction wheels along all three axes. Typical attitude sensors would be horizon scanners and a gyro-compass.

The wheel momentum storage capability sizes will be based upon fundamental considerations discussed by references [5-7]. The considerations applicable to this particular vehicle may be summarized as follows:

1. Wheel momentum storage capability should be greater than the residual vehicle angular momentum at the end of the initial acquisition sequence.

2. Wheel momentum storage capability should be greater than the maximum expected amplitude of the periodic inertial angular momentum due to disturbance torques and reference motions.

3. Wheel momentum storage capability should be sized such that the expected buildup of momentum due to secular inertial disturbance torques can be stored for a reasonable length of time to result in a long time between successive momentum corrections.

On the basis of these constraints, the wheels will be arbitrarily sized so that the secular disturbance torques will ideally require momentum corrections every two orbits in each axis. A typical secular disturbance torque acting on a vehicle would be on the order of 10^{-5} ft-lb, and would produce a momentum buildup (in inertial space) over two orbits of approximately 0.12 ft-lb-sec; therefore, each of the intermittent magnetic momentum systems will be sized to nominally remove this increment of momentum at each correction.

It can be seen from Fig. 8 and Eq. (A12) that the Optimum Intermittent scheme could perform corrections for the following conditions:

1. x wheel corrections can occur when $(\alpha - \eta) = \pi/2, 3\pi/2, \ldots$.
2. y wheel corrections can occur when $(\alpha - \eta) = \pi/2, 3\pi/2, \ldots$.
3. z wheel corrections can occur when $(\alpha - \eta) = 0, \pi, 2\pi, \ldots$.

Therefore, it is seen that even the Optimum Intermittent scheme cannot remove y wheel momentum without resulting in momentum errors coupled into the axes transverse to the correction, as well as errors in the actual correction size along the y axis. However, the utilization of the Optimum Intermittent scheme for this particular case is impractical owing to the rotation of the reference axes. This causes an inertial momentum vector to rotate as seen in the reference coordinate set; and since the rotation frequency of the momentum vector is almost identical to that of the **B** vector, the momentum will never appear in the proper wheel for removal at the same time that the minimum $(\mathbf{B} \cdot \mathbf{H}_w)$ condition occurs if the rotations are in phase. Hence, the utilization of the Intermittent scheme is implied for low-altitude orbits, as in general the difference in frequencies of the two vector rotations is on the order of ω_e , the earth's spin frequency. This would require a maximum wait of 24 hours until both the minimum condition occured and a momentum vector appeared in the proper position for removal.

The Intermittent scheme sizing will be such that 0.12 ft-lb-sec is removed at each correction. This is accomplished by use of a switch hysteresis characteristic as shown in Fig. 9. The performance of the Intermittent scheme may be determined by calculating the maximum time in each axis required to remove 0.12 ft-lb-sec, and determining the maximum momentum errors coupled into the transverse axes for each case. These errors will appear as additional disturbance torques acting on the vehicle, and will effectively reduce the off-time between successive corrections in each axis. For this particular vehicle, the

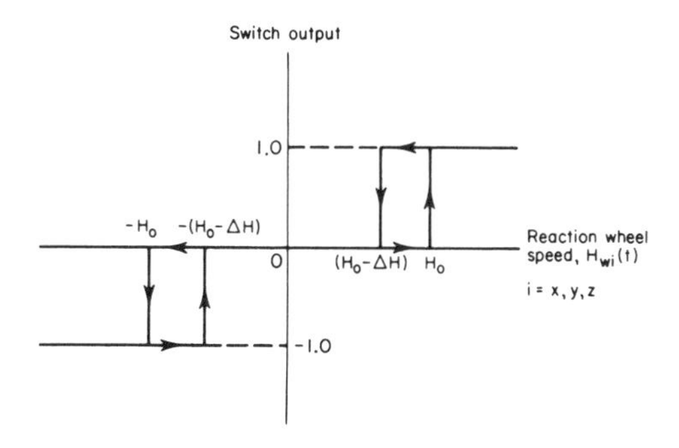

FIG. 9. Switch hysteresis characteristic for intermittent momentum dumping scheme.

magnetic system will be arbitrarily sized so that the desired correction would occur in 3 minutes at the geomagnetic equator, provided that **B** is normal to $\mathbf{H}_w$. This will require an effective current loop sizing of 400 amp-turn-ft^2 in each axis.

The results of the performance analysis for this particular example are represented in Figs. 10-13. The performance characteristics expressed for each axis are: the minimum on-time for a correction; the maximum on-time for a correction; the worst-case momentum coupling errors for a correction; and the worst-case effective off-time between successive

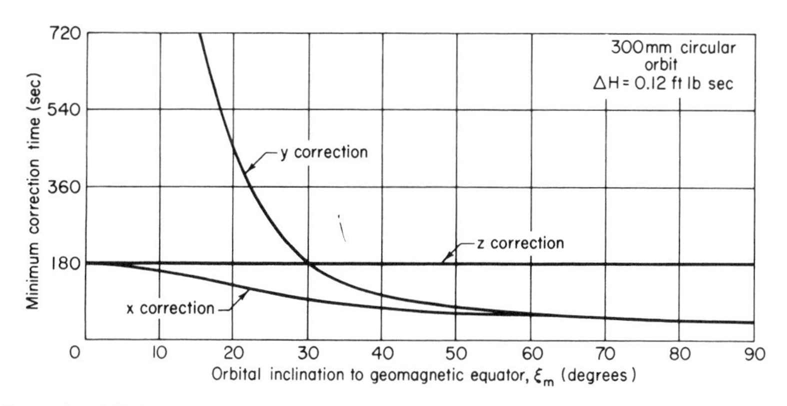

FIG. 10. Minimum on-time in each axis for intermittent scheme for various ξ_m.

corrections due to momentum coupling errors. The performance characteristics are expressed as a function of ξ_m, the instantaneous orbital inclination to the geomagnetic equator. Then, given a particular orbit, the variation of $\xi_m(t)$ over the vehicle lifetime can be estimated, and the minimum performance characteristics determined.

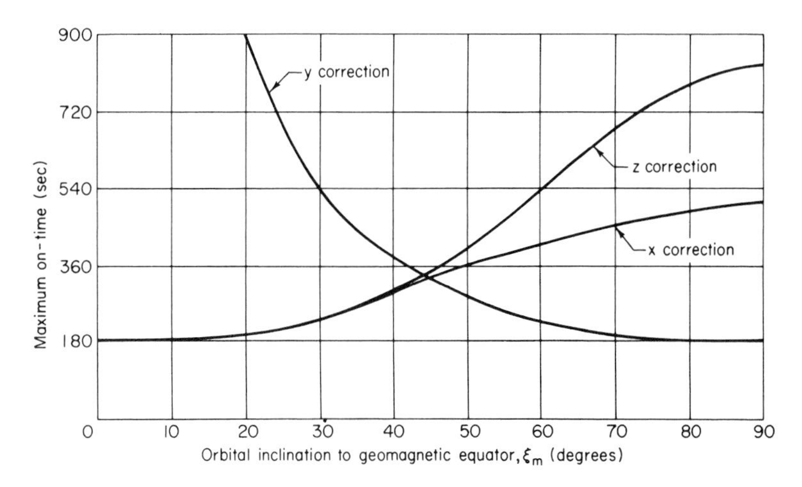

FIG. 11. Maximum on-time in each axis for intermittent scheme for various ξ_m.

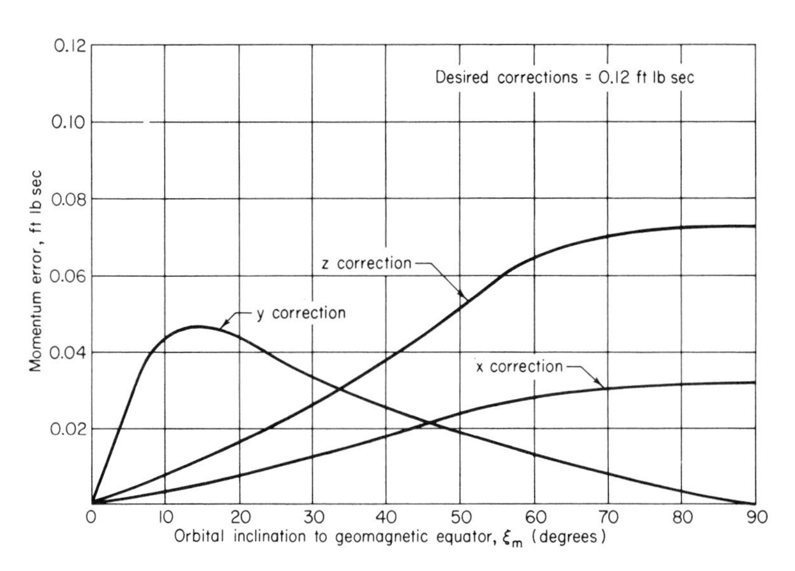

FIG. 12. Worst-case momentum errors coupled into transverse axes during one correction for intermittent scheme.

From Figs. 10 through 13, it can be seen that the momentum removal efficiency decreases as ξ_m decreases, until when $\xi_m = 0$, magnetic momentum dumping is ineffective in the y axis. This is due to the fact that the **B** vector will remain always parallel to the y axis in a geomagnetic equatorial orbit. In general, this condition can occur only for a short time in any possible satellite orbit, owing to the rotation of the geomagnetic

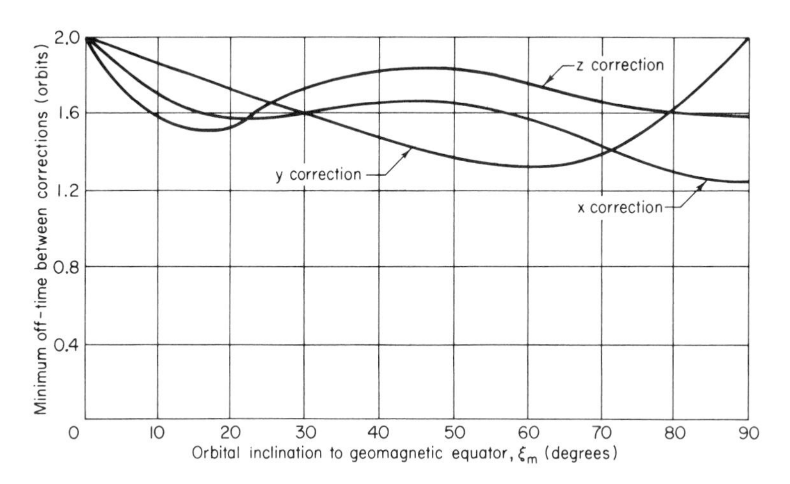

FIG. 13. Minimum off-time between successive corrections in each axis due to momentum coupling errors.

equatorial plane with the earth; and the momentum removal efficiency increases rapidly as ξ_m increases from $\xi_m = 0$, as demonstrated by the performance characteristics. For example, when $\xi_m = 30$ deg, the worst-case minimum off-time between successive corrections is reduced from two orbits to only 1.60 orbits, and the worst-case on-time for a correction is increased only from 3 minutes to 9 minutes.

This design application example has been used to illustrate the utilization of intermittent momentum removal for a satellite mission. As seen from the preceding discussion, the satellite orientation and mission requirements will greatly influence the decision as to the best scheme to utilize. However, it has been demonstrated that momentum removal utilizing the earth's magnetic field is quite applicable for many earth satellite missions, especially those with low-altitude orbits. Since the earth's magnetic field decreases rapidly with altitude, the weight and power requirements for magnetic momentum removal will increase rapidly as the orbital altitude increases.

APPENDIX
MATHEMATICAL MODEL FOR THE EARTH'S MAGNETIC FIELD

The earth's magnetic field may be approximated in the first order as the field due to a simple magnetic dipole at the center of the earth, with dipole axis inclined approximately 17 deg to the earth's equator. This approximation is sufficiently accurate for considering magnetic attitude control of an orbiting satellite [3].

The potential function for a magnetic dipole is given by:

$$\phi = -\frac{m \sin \theta}{r^2} \tag{A1}$$

where m is the dipole strength, r the magnitude of the radius vector from the dipole to the vehicle c.m., and θ is the latitudinal position relative to the magnetic equator. The magnetic field, **B**, is then the negative gradient of the potential function, ϕ:

$$\mathbf{B} = -\nabla\phi = -\frac{m}{r^3}[2 \sin \theta \hat{e}_r - \cos \theta \hat{e}_\theta], \tag{A2}$$

where $\hat{e}_r$ and $\hat{e}_\theta$ are the unit vectors directed along the vector **r** and normal to **r** in latitudinal direction. For the earth's magnetic field, the value of m is 8.1×10^{25} emu units, which gives the field in gauss for r in centimeters. The magnitude of the magnetic field, B, is given by:

$$|B| = \frac{m}{r^3}\sqrt{1 + 3 \sin^2 \theta} \tag{A3}$$

As seen from Eq. (A3) the magnetic field magnitude for a circular orbit increases as the geomagnetic latitude increases, and for a geomagnetic polar orbit, the maximum value is twice the magnitude of a geomagnetic equatorial orbit.

The position of a satellite relative to the $x_e y_e z_e$ set may be defined in terms of the geomagnetic latitude, θ, and the geomagnetic longitude, λ, as shown in Fig. 14. Then, the transformation of vectors from the $x_e y_e z_e$ set to the geomagnetic spherical set $r\,\lambda\theta$ is given by:

$$\begin{bmatrix} e_\lambda \\ e_\theta \\ e_r \end{bmatrix} = \begin{bmatrix} \cos \lambda & 0 & -\sin \lambda \\ -\sin \theta \sin \lambda & \cos \theta & -\sin \theta \cos \lambda \\ \cos \theta \sin \lambda & \sin \theta & \cos \theta \cos \lambda \end{bmatrix} \begin{bmatrix} e_{xe} \\ e_{ye} \\ e_{ze} \end{bmatrix} \tag{A4}$$

where the e_j are unit vectors along the jth axis.

The **B** field, given by Eq. (A2), may be transformed to the $x_e y_e z_e$ set by use of the inverse of (A4), resulting in:

$$\begin{bmatrix} B_{xe} \\ B_{ye} \\ B_{ze} \end{bmatrix} = -\frac{m}{r^3} \begin{bmatrix} 3 \sin\theta \sin\lambda \cos\theta \\ 3 \sin^2\theta - 1 \\ 3 \sin\theta \cos\theta \cos\lambda \end{bmatrix} \tag{A5}$$

The transformation of vectors from the $x_e y_e z_e$ set to the inertial reference set is given by:

$$\begin{bmatrix} x_i \\ y_i \\ z_i \end{bmatrix} = \begin{bmatrix} \cos(\beta_m + \omega_e t) & 0 & \sin(\beta_m + \omega_e t) \\ 0 & 1 & 0 \\ -\sin(\beta_m + \omega_e t) & 0 & \cos(\beta_m + \omega_e t) \end{bmatrix} \begin{bmatrix} \cos 17^\circ & -\sin 17^\circ & 0 \\ \sin 17^\circ & \cos 17^\circ & 0 \\ 0 & 0 & 1 \end{bmatrix} \begin{bmatrix} x_e \\ y_e \\ z_e \end{bmatrix} \tag{A6}$$

where β_m is the angle between the intersection of the geomagnetic equator with the equatorial plane and the z_i axis (line of equinoxes) at $t = 0$, and ω_e is the earth's spin rate.

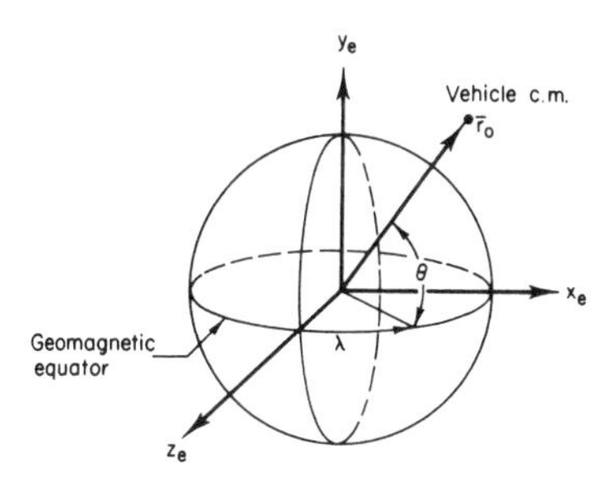

FIG. 14. Latitude and longitude measurement relative to geomagnetic equator.

The transformation of vectors from the $x_i y_i z_i$ set to the orbital reference set is given by:

$$\begin{bmatrix} x_0 \\ y_0 \\ z_0 \end{bmatrix} = \begin{bmatrix} \cos\alpha & 0 & -\sin\alpha \\ 0 & 1 & 0 \\ \sin\alpha & 0 & \cos\alpha \end{bmatrix} \begin{bmatrix} \cos\xi_e & \sin\xi_e & 0 \\ -\sin\xi_e & \cos\xi_e & 0 \\ 0 & 0 & 1 \end{bmatrix}$$
$$\times \begin{bmatrix} \cos(\beta_e + \omega_\lambda t) & 0 & -\sin(\beta_e + \omega_\lambda t) \\ 0 & 1 & 0 \\ \sin(\beta_e + \omega_\lambda t) & 0 & \cos(\beta_e + \omega_\lambda t) \end{bmatrix} \begin{bmatrix} x_i \\ y_i \\ z_i \end{bmatrix} \tag{A7}$$

Equations (A6) and (A7) may be combined to give:

$$\begin{bmatrix} x_0 \\ y_0 \\ z_0 \end{bmatrix} = \begin{bmatrix} a_{11} & a_{12} & a_{13} \\ a_{21} & a_{22} & a_{23} \\ a_{31} & a_{32} & a_{33} \end{bmatrix} \begin{bmatrix} x_e \\ y_e \\ z_e \end{bmatrix} \tag{A8}$$

where

$$
\begin{aligned}
a_{11} \quad & [\cos(\beta_m - \beta_e + \omega_e t - \omega_\lambda t) \cos 17^\circ \cos \xi_e + \sin \xi_e \sin 17^\circ] \cos \alpha \\
& + \sin(\beta_m - \beta_e + \omega_e t - \omega_\lambda t) \sin \alpha \cos 17^\circ \\
a_{12} &= [\sin \xi_e \cos 17^\circ - \sin 17^\circ \cos(\beta_m - \beta_e + \omega_e t - \omega_\lambda t) \cos \xi_e] \cos \alpha \\
& \quad - \sin(\beta_m - \beta_e + \omega_e t - \omega_\lambda t) \sin \alpha \sin 17^\circ \\
a_{13} &= \cos \xi_e \sin(\beta_m - \beta_e + \omega_e t - \omega_\lambda t) \cos \alpha \\
& \quad - \cos(\beta_m - \beta_e + \omega_e t - \omega_\lambda t) \sin \alpha \\
a_{21} &= \sin 17^\circ \cos \xi_e - \sin \xi_e \cos(\beta_m - \beta_e + \omega_e t - \omega_\lambda t) \cos 17^\circ \qquad \text{(A9)} \\
a_{22} &= \sin \xi_e \sin 17^\circ \cos(\beta_m - \beta_e + \omega_e t - \omega_\lambda t) + \cos 17^\circ \cos \xi_e \\
a_{23} &= \sin \xi_e \sin(\beta_m - \beta_e + \omega_e t - \omega_\lambda t) \\
a_{31} &= [\cos(\beta_m - \beta_e + \omega_e t - \omega_\lambda t) \cos 17^\circ \cos \xi_e + \sin \xi_e \sin 17^\circ] \sin \alpha \\
& \quad - \sin(\beta_m - \beta_e + \omega_e t - \omega_\lambda t) \cos 17^\circ \cos \alpha \\
a_{32} &= [\sin \xi_e \cos 17^\circ - \sin 17^\circ \cos(\beta_m - \beta_e + \omega_e t - \omega_\lambda t) \cos \xi_e] \sin \alpha \\
& \quad + \sin(\beta_m - \beta_e + \omega_e t - \omega_\lambda t) \sin 17^\circ \cos \alpha \\
a_{33} &= \cos \xi_e \sin(\beta_m - \beta_e + \omega_e t - \omega_\lambda t) \sin \alpha \\
& \quad + \cos(\beta_m - \beta_e + \omega_e t - \omega_\lambda t) \cos \alpha
\end{aligned}
$$

The use of either Eq. (A4) or (A8) can be used to write the components of **B** at the satellite position in orbit; therefore, these equations can also be used to write λ and θ as functions of β_m, β_e, $\omega_e t$, $\omega_\lambda t$, and ξ_e. The vectors $\hat{e}_r$ and $\hat{e}_{z0}$ must be coincident if both equations are to represent the satellite position in orbit; hence,

$$
\begin{aligned}
a_{31} &= \cos \theta \sin \lambda \\
a_{32} &= \sin \theta \qquad \text{(A10)} \\
a_{33} &= \cos \theta \cos \lambda
\end{aligned}
$$

where the a_{ij} are given by Eq. (A9). The components of **B** can then be written in the $x_0 y_0 z_0$ set solely as functions of the parameters β_m, β_e, $\omega_e t$, $\omega_\lambda t$, and ξ_e:

$$
\begin{bmatrix} B_{x0} \\ B_{y0} \\ B_{z0} \end{bmatrix} = -\frac{m}{r^3} \begin{bmatrix} 3a_{32}a_{31}a_{11} + 3a_{12}a_{32}^2 - a_{12} + 3a_{13}a_{32}a_{33} \\ 3a_{32}a_{31}a_{21} + 3a_{22}a_{32}^2 - a_{22} + 3a_{23}a_{32}a_{33} \\ 3a_{32}a_{31}^2 + 3a_{32}^3 - a_{32} + 3a_{32}a_{33}^2 \end{bmatrix} \frac{m}{r^3} \begin{bmatrix} a_{12} \\ a_{22} \\ -2a_{32} \end{bmatrix} \qquad \text{(A11)}
$$

where the a_{ij} are given by Eq. (A9).

In the general case, Eqs. (A11) are complicated functions of time; however, for several special cases, simplifications can be made. These special cases are: (1) orbits in which ω_0 (orbital rate) is much greater than ω_e, resulting in simple expressions for **B**, if only several orbits are considered; and (2) orbits which lie in the earth's equatorial plane. For both cases, Eqs. (A11) reduce to:

$$\begin{bmatrix} B_{x0} \\ B_{y0} \\ B_{z0} \end{bmatrix} = \frac{m}{r^3} \begin{bmatrix} \cos(\alpha - \eta_m) \sin \xi_m \\ \cos \xi_m \\ -2 \sin(\alpha - \eta_m) \sin \xi_m \end{bmatrix} \tag{A12}$$

where η_m is a phase angle measured from the ascending node of the orbit relative to the earth's equator to the ascending node of the orbit relative to the geomagnetic equator, and ξ_m is the instantaneous inclination of the orbit plane to the geomagnetic equator. In general, both η_m and ξ_m are time-varying functions, and can be assumed constant over a small number of orbits only if $\omega_0 \gg \omega_e$. However, for an equatorial orbit, ξ_m is constant (17 deg and η_m is just given by:

$$\eta_{eq} = \omega_e t \tag{A13}$$

In general, the angles ξ_m and η_m are given by:

$$\begin{aligned} \xi_m = \cos^{-1} [&\cos \xi_e \cos 17^\circ \\ &+ \sin \xi_e \sin 17^\circ \cos(\beta_e - \beta_m - \omega_e t + \omega_\lambda t)] \end{aligned} \tag{A14}$$

$$\eta_m = \sin^{-1} \left[\frac{\sin 17^\circ \sin(\beta_e - \beta_m - \omega_e t + \omega_\lambda t)}{\sin \xi_m} \right] \tag{A15}$$

Therefore, as seen from Eqs. (A14) and (A15), the angles η_m and ξ_m can be assumed constant over several orbits only if $\omega_0 \gg \omega_e$.

If the desired orientation of the vehicle body axes is not coincident with the $x_0 y_0 z_0$ coordinate set, the components of **B** given by Eq. (A12) can be resolved to any other reference set by use of the general matrix transformation $[A]$, given by Eqs. (1) and (2).

References

1. J. S. White, F. H. Shigimoto, and K. Bourgum, Satellite Attitude Control Utilizing the Earth's Magnetic Field, *NASA Tech. Note D-1068*, August 1961.
2. A. G. Buckingham, A new method of attitude control utilizing the earth's magnetic field for long life space vehicles, *ARS Guidance and Control Conf.*, Stanford, California, August *7-9, 1961.*

3. R. E. Fischell, Magnetic and Gravity Attitude Stabilization of Earth Satellites, *Johns Hopkins Univ. Appl. Phy . Lab. Rept. CM-996*, May 1961.
4. E. H. Vestine, The Geomagnetic Field, Its Description and Analysis, *Carnegie Inst. of Washington Publ. 480*, 1947.
5. R. J. McElvain, Effects of Solar Radiation Pressure upon Satellite Attitude Control, *ARS Guidance and Control Conf., Stanford, California, August 7-9, 1961.*
6. R. E. Mortensen, Design considerations of inertia wheel systems for attitude control of satellite vehicles, *Joint Automatic Control Conf., Boulder, Colorado, June 1961.*
7. R. K. Whitford, Design of attitude Control Systems for Earth Satellites, *Space Technol. Labs. Rept. 2313-0001-RU-000*, June 1961.

Torques and Attitude Sensing in Spin-Stabilized Synchronous Satellites

DONALD D. WILLIAMS

System Design Department, Project Syncom, Hughes Aircraft Company, Culver City, California

I. INTRODUCTION

THE SPIN-STABILIZED 24-HOUR or synchronous communication satellite was first proposed by the Hughes Aircraft Company in the fall of 1959. The use of spin stabilization in this orbit, as contrasted with three-axis attitude control, was seen as a means of achieving this difficult but extremely desirable orbit at an early date with existing boosters of relatively low cost. After much study, the concept proposed by Hughes was accepted, with essentially no modification in the attitude control and station-keeping concepts of the original design, by the National Aeronautics and Space Administration, in the form of Project Syncom.

Syncom Mk I will be the first attempt to achieve the 24-hour orbit. The satellite will be launched from AMR by Thor-Delta vehicle, and no attempt will be made to eliminate the 33-deg inclination resulting from the latitude of the Cape and the azimuth required for reasons of range safety. Figure 1 is a photograph of a satellite prototype built on Company funds by Hughes prior to receiving the Syncom contract; the Syncom satellite is very similar in appearance. The diameter is 28 inches, permitting use of the low-drag nose fairing of the Delta.

In order to understand the attitude sensing and control concept of Syncom, it is first necessary to describe the method of launch and subsequent operation. The boost vehicle injects the satellite at or near the perigee of a transfer ellipse. At or near the apogee of this transfer orbit,

FIG. 1. Spin-stabilized synchronous satellite prototype.

when the satellite has reached the synchronous, circular orbital radius of 22,752 nautical miles, the final boost is applied. The solid-propellant apogee motor required for this boost is a spherical motor within the Syncom satellite itself. Since during the transfer orbit the inertial velocity vector of the satellite rotates essentially 180 deg, and since the spin axis of the satellite and third-stage motor is nearly parallel to the velocity at injection into the transfer orbit, the apogee motor appears, when the satellite is on the vehicle, to be a retrorocket. The spin is initially imparted through the third stage by the spin-up rockets as in the case of all Delta payloads.

In order to perform a significant communication function in the high-altitude orbit, the satellite transponder transmitter must have a certain degree of antenna directivity. At this altitude, a pencil beam 17-deg wide will encompass the earth and provide about 17 db gain. A "pancake" beam, symmetric around the axis of spin, will encompass the earth if it is 17 deg in thickness, and will give a gain of about 8 db. The "pancake" beam has been selected for Syncom as the simplest technique compatible with spin stabilization. The pancake beam is useful only when the earth is in the plane of the pancake; therefore to obtain the maximum gain continuously the spin axis must be normal to the plane of the orbit. Since, as can be understood from the foregoing description of the injection process, the spin axis is initially in the plane of the orbit, there is a requirement to reorient the spin axis through 90 deg.

The solar power supply for Syncom is a cylindrical array around the spin axis. For optimum performance, such an array should be illuminated nearly at normal incidence to the spin axis. By placing the spin axis normal to the plane of the orbit, we achieve a variation between the sun line and the normal which, in the course of a year, is equal to plus and minus the inclination angle between the orbit plane and the ecliptic plane. For our initial launch, we can have this angle as small as 10 deg or up to 25 deg, depending on where in our launch "window" we actually launch. When we eventually achieve the equatorial, truly stationary orbit with this type of satellite, the variation will be $\pm$ 23.4 deg.

In order to reorient the satellite, we incorporate a pulsed jet precession system in the spacecraft. Two titanium spheres store 2.5 lb of nitrogen gas which supplies the precession jet. The geometry of the jet can be seen in Fig. 2; its thrust is parallel to the axis of spin, and the jet is offset by the maximum possible radius, approximately 13 inches. The jet is pulsed in synchronism with the spin of the satellite, producing a net average precession torque in a fixed direction in space.

The same concept of a pulsed jet can also be applied for velocity control or "station keeping." The velocity control jet, also shown in

Fig. 2, thrusts normal to the spin axis and through the center of gravity. When it is pulsed in synchronism with the spin, a net acceleration of the spacecraft is produced. When the spin axis is in its final attitude normal to the orbital plane, this provides the capability of control of the orbital velocity and therefore the period.

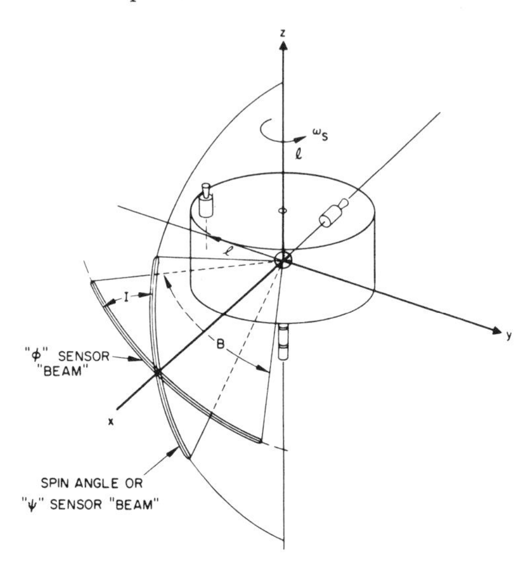

FIG. 2. Geometry of the attitude sensing and control system.

In Syncom Mk 1, there are actually two pairs of jets, both with the basic geometry described above. The second pair, however, uses hydrogen peroxide rather than stored gas. Either pair may be used in the manner described above; however, the primary function of the peroxide system is based on continuous rather than pulsed operation; before the satellite is reoriented, the axial peroxide jet may be used to control the period if it is operated at the correct time in the orbit. Either sense of control is possible from the single jet, depending on the choice of the time, since the inertial velocity is rotating relative to the thrust axis once per orbit (or, per day). Finally, we may observe that after reorientation the axial jet may be used to control inclination, an important feature in more advanced systems where this control will permit use of fixed ground antenna reflectors.

The basic advantages of spin stabilization for the synchronous orbit are therefore not limited to light weight and elimination of a multiplicity of attitude control jets, reaction wheels, gyroscopes, and other equipment; spin stabilization also permits complete control of the orbit with only two jets, leaving room for redundancy in the control system. We believe that spin stabilization is not a temporary expedient, but rather the "ultimate" answer in that we can foresee no future communication satellite requirements for which three-axis attitude stabilization is necessary. We are already working on a more advanced design in which a pencil beam will be "despun" to continually illuminate the earth, eliminating the 9-db penalty we pay with the simple initial design. This despinning may be done entirely electronically, or by a single rotating antenna assembly whose bearings need not be exposed to vacuum.

Having justified our interest in spin stabilization in the synchronous orbit, we now proceed to analyze some of its features more quantitatively.

II. Attitude Sensing

In the type of spacecraft we have described, attitude sensing is necessary in order to determine the correction required to obtain the desired spin axis pointing direction; it is also required in order to pulse the jets at the proper phase in the spin cycle to make the desired correction to attitude or velocity.

Syncom provides for two types of attitude sensing—solar and radio frequency. Solar sensing is the primary source of attitude information. Radio frequency attitude sensing is accomplished entirely by ground equipment and does not affect the spacecraft design.

Figure 2 illustrates the geometry of the Syncom solar sensing system. The solar sensors use simple slit optics and are shown assembled and disassembled in Fig. 3. The two sheet aluminum cups are held apart by five spacers, and their edges form, in effect, two slits about 1-inch apart and 0.008 to 0.012-inch wide in front of the two parallel-connected

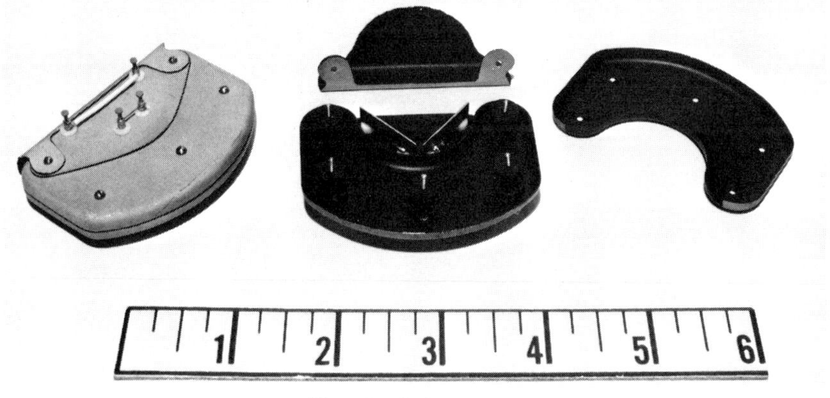

Fig. 3. Solar sensors.

solar cells. When the sun line is in the plane of the slits, maximum output is obtained. The beam width normal to the plane of the slits is about 0.7 deg; in the plane of the slits, useful output is obtained out to $\pm$ 80 deg from the plane of symmetry. The beam shape is, therefore, a wide thin fan.

Figure 2 shows the geometry of the solar sensors, which are termed "ψ" and "ψ_2." The "ψ" sensor is a spin angle sensor; its telemetered output pulse serves as a reference for spin angle and occurs when the sun line is in the (X, Z) plane. The "ψ_2" sensor is rotated about the X axis relative to the "ψ" sensor through the angle I, which has been chosen as 35 deg. The pair of sensors therefore forms a "V-beam" system, and the relative time between their output pulses is used to measure the angle between the sun line and the spin axis. Quantitatively, let ϕ = angle between sun line and spin axis and ψ_2 = angle of rotation from ψ sensor maximum to ψ_2 sensor maximum; then

$$\cot \phi = \sin \psi_2 \cot I \tag{1}$$

On the ground, a rotating drum contact device (Fig. 4) is used to measure the attitude and control the pulses. The drum is synchronized so that a reference pulse, generated when a fixed contact touches the cam, always coincides with the reception of the ψ solar sensor pulse by the

FIG. 4. Synchronous control drum unit.

ground station. Rotating the left-hand dial to the correct angle gives the required pulse phasing, the command pulse being generated by the movable contacts attached to the dial. To compensate for propagation time, provision is made for transponding the command pulse to establish a correction constant for the dial setting. The right-hand dial is rotated until its contacts generate a pulse in coincidence with that received from the ψ_2 solar sensor. The angle ψ_2 may then be read from the scale engraved next to the frame. The second set of graduations on the dial are engraved according to Eq. (1) to give ϕ directly.

The complete solar sensing system in Syncom includes redundant pairs of ψ and ψ_2 sensors, and in addition three more "ψ-type" quadrant sensors disposed about the roll axis 90-deg apart. These sensors plus the primary ψ sensors define four quadrants. In an alternate mode of control, the jets may be turned on while the sun line is in the selected spin quadrant, rather than by use of the drum device.

It is clear that solar sensing alone can give only two of the three attitude angles necessary to completely define the geometric situation. For this reason, we have rf sensing by the ground as a back-up. Radio frequency sensing may take two forms. By making a record of the satellite signal strength over a full day in orbit, it is possible to detect and estimate the magnitude and direction of the attitude error. This method alone would be good insurance against "getting lost," for the requirement for correct spin axis attitude stems primarily from consideration of signal strength. In addition to the signal strength method, we are also providing for measurement of the polarization of the signal received from the satellite transponder. The transponder transmitting antenna is polarized along the spin axis. At the frequency of approximately 1800 mc which is employed, Faraday rotation is not expected to constitute a problem. Use of the polarization information plus solar sensing to completely determine the attitude in space involves an interesting but somewhat lengthy exercise in spherical trigonometry.

The method of computing the desired direction of precession is of some interest. We can compute the desired final angle between the spin axis and the sun line; one method of doing this is to find, from the orbital inclination and longitude of the ascending node, the declination and right ascension (R.A.) of the pole of the orbit. If we let

$$t = (\text{R.A. of sun}) - (\text{R.A. of pole of orbit})$$

and consider a geographic position at a latitude equal to the sun's declination, we may compute using one of the rapid methods developed for celestial navigation the "altitude" of the orbit normal considering

t as the meridian angle. The desired value of ϕ after precession is 90 deg minus the computed altitude. We may make a similar computation for the initial attitude based on the expected injection conditions. This method also gives another important parameter, $\Delta\theta$, which is defined as the angle of rotation about the sun line between the initial and final planes defined by the spin axis and the sun line. $\Delta\theta$ is simply the difference in azimuth angles obtained for the spin axis directions at the subsolar point.

There is an interesting analogy between the geometry of precession for the satellite and Mercator navigation. Since the reference direction is the sun line, we imagine it to correspond to the north pole of a Mercator projection. The angle ϕ then corresponds to colatitude and the angle $\Delta\theta$ corresponds to difference in longitude. When the satellite is precessed, the "tip" of a unit vector along the spin axis moves, on the average, in the direction of the average applied torque, whose angle relative to the plane of the spin axis and sun line is controlled. This angle between the torque and the spin axis-sun line plane is, in the Mercator plot, the angle between the direction of motion of the spin axis and the meridian. If the dial setting of the control unit is not changed during the precession, this "course angle" is constant and the spin axis traverses a straight line—a rhumb line—on the Mercator chart. Therefore we have a simple graphic solution of the precession problem. Without this, use of the quadrant mode of control would present a complex computation problem.

One further point of analogy is of interest. Absence of additional information over and above that provided by solar sensing corresponds to the situation in which a navigator is unable to determine longitude. Rather than resort to "square sailing" as practiced before the days of chonometers at sea, we solve this problem by precessing in the computed direction until the correct "latitude" is reached. It is obvious that when the "course angle" approaches 90 deg this becomes inaccurate. This is one consideration in selecting the most desirable launch time on any given date and also in deciding whether to orient the spin axis parallel or antiparallel to the orbital angular momentum.

III. Nutation

At the outset, we wish to distinguish between nutation and precession; we define "nutation" as the motion of a body which exists in the absence of disturbing torques but deviates from uniform spin about a fixed axis; and "precession" as the motion which results from an applied torque. This distinction is made to prevent confusion, since the free-body, torque-free motion has sometimes been termed precession.

A considerable economy in analysis results from assuming that the principal moments of inertia of a spin-stabilized spacecraft about axes normal to the spin axis are equal; this condition is met by Syncom within a maximum of about 7% owing primarily to the different masses of peroxide and cold gas supplies located at the ends of two diameters 90 deg apart. For a symmetric body, the Euler equations [1] become

$$I_x \dot{\omega}_x = (I_x - I_z)\, \omega_z \omega_y + N_x \tag{2}$$

$$I_x \dot{\omega}_y = -(I_x - I_z)\, \omega_z \omega_x + N_y \tag{3}$$

$$I_z \dot{\omega}_z = N_z \tag{4}$$

and admit the solution

$$\omega_x = A \sin \Omega t \tag{5}$$

$$\omega_y = A \cos \Omega t \tag{6}$$

$$\Omega = \frac{(I_x - I_z)}{I_x}\, \omega_z \tag{7}$$

in the case for which the torques N are zero. Here z denotes the spin axis or axis of symmetry; the (x, y, z) axes form a right-handed system of body axes, and the ω's are components of the instantaneous angular velocity vector. This motion is the characteristic free-body nutation; it corresponds to motion of the geometric spin or z axis in a cone of half-angle θ_n at angular frequency ω_n, where

$$\theta_n \cong A/\omega_n \tag{8}$$

$$\omega_n = I_z \omega_z / I_x \tag{9}$$

Nutation is excited in Syncom by separation disturbances, apogee motor thrust eccentricity, and the operation of the precession jet. Excessive nutation would cause errors or at least "jitter" in the solar sensing system; therefore it is desirable to damp the nutation artificially and to predict the rate of decay.

The physical principles underlying nutation damping are brought out by considering the relations between the energy and the angular momentum. In the constant-amplitude nutation, both the total angular momentum and the kinetic energy are constant. They may be expressed as

$$L^2 = I_z^2 \omega_z^2 + I_x^2 A^2 \tag{10}$$

$$E = \tfrac{1}{2}(I_z \omega_z^2 + I_x A^2) \tag{11}$$

By combining Eq. (10) and (11), we find

$$E = \frac{1}{2I_z}(L^2 + (I_z - I_x) I_x A^2) \tag{12}$$

Any motion of passive parts within the satellite will not change the angular momentum, but may decrease the kinetic energy owing to frictional losses. Therefore we arrive at the conclusion that for $I_z > I_x$ nutation represents excess energy at constant angular momentum; if $I_z < I_x$ nutation represents a loss in energy at constant angular momentum and such a configuration is unstable and can only be damped by active devices.

In considering a system initially nutating with amplitude A, which damps to zero, we find that the final spin speed is given by

$$\omega_{zf}^2 = \omega_{zi}^2 + \frac{I_x^2}{I_z^2} A_i^2 \tag{13}$$

To estimate the rate of decay of nutation, we consider the damper as a perturbation and differentiate Eq. (12), obtaining

$$\frac{(I_z - I_x) I_x}{I_z} A \frac{dA}{dt} = -P \tag{14}$$

where P is the power dissipated by the damper.

The cause of the power dissipation will be the time-varying acceleration which is experienced at a general point attached to a nutating body. In vector form, the motion of a small mass m constrained to the body by a force $\mathbf{F}$ is given by

$$\mathbf{F} + m\delta\mathbf{r}'' + m[\boldsymbol{\omega} \times (\boldsymbol{\omega} \times \delta\mathbf{r}) + \dot{\boldsymbol{\omega}} \times \delta\mathbf{r} + 2\boldsymbol{\omega} \times \delta\mathbf{r}'] = -m[\boldsymbol{\omega} \times (\boldsymbol{\omega} \times \mathbf{r}_0) + \dot{\boldsymbol{\omega}} \times \mathbf{r}_0] = -m\,\mathbf{a}_d \tag{15}$$

Here $\delta\mathbf{r}$ is the displacement of the mass relative to the rest position $\mathbf{r}_0$, and primes denote a time-derivative vector whose components are the derivatives of the body axis components of the original vector. This expression, when expanded, gives the following useful set of equations:

$$\frac{F_x}{m} + \delta\ddot{x} - \omega_z^2\delta x - 2\omega_z\delta\dot{y} = +\omega_z^2 x_0 - \omega_n z_0 A \sin \Omega t \tag{16}$$

$$\frac{F_y}{m} + \delta\ddot{y} - \omega_z^2\delta y + 2\omega_z\delta\dot{x} = +\omega_z^2 y_0 - \omega_n z_0 A \cos \Omega t \tag{17}$$

$$\frac{F_z}{m} + \delta\ddot{z} = -(\omega_z + \Omega) A [x_0 \sin \Omega t + y_0 \cos \Omega t] \tag{18}$$

The terms on the right are "driving accelerations" for the damper. It is seen that a damper on the spin axis should be displaced as far from the center of gravity as possible and will move laterally, while a damper at $z = 0$ should be placed at a maximum radius and will move along the spin axis.

In Syncom, the nutation damper is based on the latter geometry, space near the axis being hard to obtain. The damper is a simple tube, about $\frac{3}{8}$ inch in diameter and 7 inches long, roughly 30% filled with mercury. The axis of the tube is parallel to the spin axis. Because of space limitations, the device is resonant somewhat above the frequency Ω to be expected. The resonant frequency is proportional to the spin speed. The weight of the device has been limited to 0.1 lb total.

Prediction of the time constant involves some interesting dynamic similitude relations, since the damper is obviously g sensitive and a dynamic model test in the atmosphere and in the presence of gravity is likely to be misleading. To obtain performance predictions, we have used the technique of placing the damper on a pendulum with a bifilar support to maintain the tube axis horizontal. The pendulum frequency is scaled to account for the difference between the centripetal acceleration experienced in the satellite and the one g field in the experiment. The scaling is complicated by the high surface tension of mercury, which requires dimensional scaling of the damper by the square root of the acceleration ratio, a scale-up of about 3 to 1. This requirement precludes our obtaining constant Reynolds number. The predicted damping time constant is about 2 minutes. The damper has no caging provision and will destabilize the assembly of third stage and payload; the predicted time constant of divergence is somewhat over 1000 seconds. This gives us a comfortable margin but gives us reservations about trying to improve the damper effectiveness.

It is interesting to note that similar experiments show that the effect of nearly 5 lb of hydrogen peroxide with free surfaces in spherical containers is essentially negligible. Concern with this problem led us to make a pendulum experiment; we were unable to measure any effect, although the scaling was such that a predicted divergence time constant of 10 days would have been apparent.

IV. Effects of Jet Pulsing

In the Syncom spacecraft, the attitude jet is pulsed on for 60 deg of the spin cycle, or for 90 deg in the quadrant mode. When the jet is pulsed, it creates a torque around a fixed body axis which contains a constant term, a component at the spin frequency, and a series of terms at harmonics of the spin frequency. Analysis of the problem is facilitated by utilizing the complex representation of the angular velocity in body axes:

$$\omega = \omega_x + j\omega_y \tag{19}$$

$$N = N_x + jN_y \tag{20}$$

$$\dot{\omega} = -j\Omega\omega_z + (N/I_x)$$

and the equation of transformation to space coordinates

$$j\dot{\eta} = \omega \exp(j\omega_z t) \tag{21}$$

where the complex vector η represents the (small) deviation of the spin axis from its reference direction. The pulsed torque may be expanded in complex Fourier series and the results lead to a Fourier series expression for η which contains the steady precession term due to the spin-frequency component of the torque; and in addition a set of terms at the spin frequency and its harmonics which represent a deviation from uniform precession. For a typical satellite design, the moment-of-inertia ratio I_z/I_x is between 1.3 and 1.4; for such a spacecraft, summation of the series to find the mean-square angular deviation from uniform precession shows that this deviation is within 1% or less of the value by the average body-fixed torque component alone. This component is given by the simple unbalance relation

$$|\eta| = \frac{|\mathbf{N}|}{(I_z - I_x)\,\omega_z^2} \tag{22}$$

For Syncom its value, for the initial jet thrust of 1 lb, will be about 2 milliradians at nominal spin speed of 165 rpm.

In addition to the particular solution which results from solution of the Euler equations with pulsed driving terms, there is a free-body nutational motion required to satisfy the initial conditions when the pulse train is started and stopped. The residual nutation term is of nearly the same magnitude as the driven term resulting from the series expansion.

The fact that oscillations of the spin axis do not tend to build up during pulsed-jet precession is of considerable interest. During a period 2 years ago when certain high-ranking scientists were skeptical concerning our ability to apply the laws of gyroscopic motion in this manner, we constructed a three-degree-of-freedom dynamic model representing the satellite. This model, mounted on an ingenious guidance head bearing borrowed from NOTS, Inyokern, was spun and precessed by command and provided a convincing demonstration that such a body could be controlled and would not "tumble."

V. Disturbing Torques

Even a small satellite such as Syncom spinning at 2.7 revolutions per second is a rather impressive gyroscope. At a typical roll moment of inertia of 1.4 slug ft^2, the angular momentum of Syncom is 3.3×10^8 cgs units (dyne-cm-sec). Several disturbing effects, however, can be important for such a satellite; they include gravity gradient, magnetic effects, and solar radiation pressure unbalance.

In regard to gravity-gradient torques, we first observe that, since the spin axis is normal to the plane of the orbit, the dominant inverse-squared attraction does not give rise to any torque on the satellite. Therefore any such effect will be associated with oblateness. The torque due to oblateness would also vanish if we were in the equatorial plane; however, the initial Syncom satellite will be in an inclined orbit.

To estimate the order of magnitude of gravity-gradient precession rate, we first observe that in the inverse square field this rate is given by the expression

$$\dot{\psi} = 3g_0 \frac{r_e^2}{r_0^3} \sin\theta \cos\theta \frac{(I_z - I_x)}{I_z \omega_z} \tag{23}$$

Here g_0 is surface gravity at the earth's radius r_e; r_0 is the orbital radius and θ the angle between the radius from the earth's center and the spin axis. The maximum precession rate due to oblateness would be of the order of magnitude of

$$\frac{g_0 r_e^4}{\omega_z r_0^5} J, \qquad J = 0.001637 \tag{24}$$

which gives a value of 0.36×10^{-6} radians per year.

The presence of a magnetic field, inducing eddy currents in the satellite, has two effects: the spin is damped, and the spin axis precesses toward the instantaneous magnetic field. Ignoring inductance of the structure and assuming a reasonable degree of symmetry so that the current can be assumed to flow in simple planar distributions, it is possible to show that (in a constant magnetic field)

$$\frac{d\theta}{dt} = -\frac{\sin\theta\cos\theta}{\tau_0} \tag{25}$$

$$\frac{d\omega_z}{dt} = -\frac{\sin^2\theta}{\tau_0} \tag{26}$$

where θ is the angle between the spin axis and the magnetic field. We have in the case of a field B_0 normal to the axis

$$I_z \omega_z \dot{\omega}_z = -P_0 \tag{27}$$

$$\tau_0 = -\frac{\omega_z}{\dot{\omega}_z} = \frac{I_z \omega_z^2}{P_0} \tag{28}$$

Therefore precession rate can be estimated if the power P_0 dissipation by eddy currents in a normal magnetic field can be calculated.

During the preparation of a proposal for a spinning low-altitude orbiter, the author was required to make fairly exact estimates of the damping time constant considering the proposed sheet-metal members in some detail. This is much more important for low-altitude designs because of the inverse sixth power relationship between damping and radius. A very simple upper bound can be derived for this effect which is applicable directly to Syncom, however. It would be difficult to conceive of a more rapidly damped configuration that a thin conducting cylinder with two, massless, perfectly conducting end planes forming short circuits across the ends. All of the mass of the cylinder is concentrated in the sides. Independent of the dimensions of such a cylinder, it is found that

$$\tau_0 = \frac{2\rho}{\sigma B_0^2}. \tag{29}$$

The radial component of the earth's field is the most significant because of the geometry of the spin axis relative to the orbit. Since Syncom will not exceed a magnetic latitude of 50 deg (33 deg + 17 deg) this radial component [2] will not exceed 1.67×10^{-7} webers per square meter at the orbital radius. Assuming an aluminum cylinder, we obtain a minimum damping time constant of 170 years. For the equatorial orbit, the damping and precession will be even less rapid.

Hysteresis effects may contribute something to the damping of spin; however, we are not attempting to make any predictions, as this is difficult even for low orbiters in fields 300 times as intense.

Precession due to magnetic moment of the satellite is not expected to be a serious problem; with the maximum normal field, precession rate is approximately 5×10^{-9} rad/sec/amp-turn-meter^2. The inappropriate character of magnetic torque coils [4] for this application is easily demonstrated, especially when we consider the equatorial orbit.

Solar radiation pressure unbalance is the most significant torque we have estimated. A very simple model for calculation of this effect is a cylinder with perfectly reflecting ends and completely absorbing sides. For such a body, the torque is

$$N = \frac{\pi}{4} \frac{S}{c} a A_p \cos \delta \sin \delta \tag{30}$$

where S = solar constant
c = velocity of light
a = radius of cylinder
$A_p = 2ah$ = projected area, where h is the height of the cylinder
δ = angle between the sun line and the normal to the axis
$S/c = 0.957 \times 10^{-7}$ lb/ft^2.

The assumption of absorbing ends and reflecting sides gives the same result with a change in sign.

Consideration of the varying geometry during the course of a year shows that there is an average precession of the spin axis in the direction of the intersection of the orbital plane (normal to the initial spin axis) and the ecliptic plane; by integration of the precession rate, we find that its magnitude is

$$\dot{\psi} = \frac{\pi}{16} \frac{S}{c} a A_p \frac{\sin 2i}{I_z \omega_z} \tag{31}$$

where i is the inclination of the orbit to the ecliptic, a maximum of 25 deg. The computed maximum disturbance of the spin axis for Syncom is 3.7 deg per year, an upper bound based on assuming the maximum possible unbalance of the satellite surfaces.

One further effect is present in the inclined orbit; it is not a precession of the spin axis in inertial space, but rather the effect due to regression of the nodes in the inclined orbit, which is at a rate of approximately 4 deg per year. This will require roughly a 2-deg per year correction of the spin axis to maintain it normal to the orbital plane.

VI. Conclusions

We believe that disturbances of the satellite spin axis attitude and spin rate in the synchronous orbit are essentially negligible; furthermore, we are equipped to exercise control over the attitude throughout the life of the satellite. We feel that spin stabilization has a considerably wider range of practical application that is ordinarily recognized. We have shown that complete velocity control in orbit is possible with fewer components in a spinning spacecraft than in a nonspinning one and indicated that precise determination of the attitude is possible. We have also pointed out that key portions of a spacecraft may be despun if necessary, in some cases with far less complication than if three-axis attitude control were substituted.

With regard to the effect of torques, I would like to call attention to a remark made by Dr. Harold Rosen, who can be considered the inventor of the spinning synchronous communication satelite. If we consider two spacecraft of roughly similar size and weight, one spinning and the other nonspinning, then a constant torque acting over a time interval will produce certain angular displacements of both spacecraft. Under these circumstances, the angular displacement of the nonspinning spacecraft will be of the order of magnitude of the displacement of the spinning one, multiplied by the angle in radians through which the spinning craft rotates about its axis during the time interval. This observation brings out in an interesting way the advantages of spin stabilization.

References

1. Goldstein, "Classical Mechanics." Addison-Wesley, Readings, Massachusetts.
2. C. W. Allen, "Astrophysical Quantities," Oxford Univ. Press (Athlone), London and New York.
3. W. R. Bandeen and W. P. Manger, Angular Motion of the Spin Axis of the Tiros I Meteorological Satellite, *NASA Tech. Note D-571*, April 1961.
4. L. H. Grasshoff, A method for Controlling the Attitude of a Spin-Stabilized Satellite, *ARS Preprint No. 1501-60*, December 1960.
5. D. D. Williams, Dynamical Analysis and Design of the Synchronous Communication Satellite, *Hughes Aircraft Company Technical Memorandum No. 649*, May 1960.

Note Added in Proof

The first successful Syncom spacecraft was launched on July 26, 1963. On the following day, the axial hydrogen peroxide jet was used to control the period of the orbit, and on July 31 the spin axis was successfully precessed by pulsed operation of the same jet. Subsequently, both lateral peroxide and nitrogen jets were pulsed in a series of maneuvers which resulted in essentially perfect synchronism of the orbit with the earth's rotation. On August 18, a small correction of the attitude was made with nitrogen using the quadrant mode. No further control was necessary until November 27, when the lateral hydrogen jet was used to make a correction to the period and simultaneous reduce the eccentricity of the orbit to a value less than 0.0001. Operation of the attitude sensing and control systems has been satisfactory in all respects.

On the Motion of Explorer XI around Its Center of Mass[1]

G. COLOMBO[2]

Smithsonian Institution Astrophysical Observatory, Cambridge, Massachusetts

I. OBSERVED MOTION AND ITS POSSIBLE EXPLANATION

THE RIGID-BODY MOTION of Explorer XI (Fig. 1) around its center of mass can, after the complete damping of the spinning motion, be

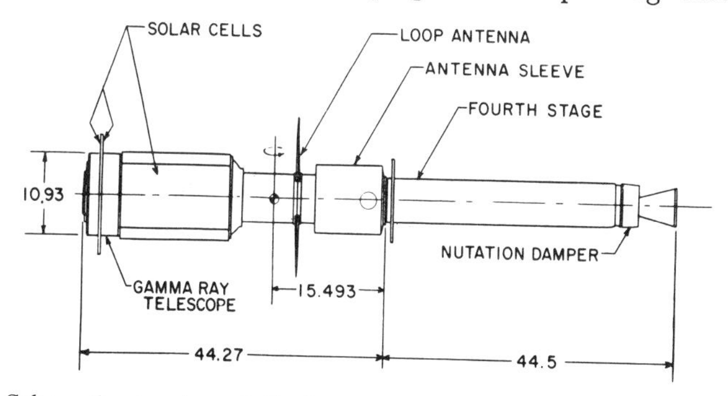

FIG. 1. Schematic drawing of Explorer XI. The total weight is 95.114 lb. The moments of inertia are $A_{\max} = 143.792$ in-lb-sec^2; $A_{\min} = 143.601$ in-lb-sec^2; and $C = 3.405$ in-lb-sec^2.

[1] This work was supported by a grant from National Aeronautics and Space Administration.

[2] On leave of absence from the University of Padua, Italy.

represented as a rotational motion around an axis defined by the vector $\boldsymbol{\Omega}$, normal to the figure axis Z of the satellite; the $\boldsymbol{\Omega}$ axis is perturbed by several effects. The observational material [1, 2] clearly indicates that the motion of $\boldsymbol{\Omega}$ is precessional (see Fig. 2). In addition, within the limit

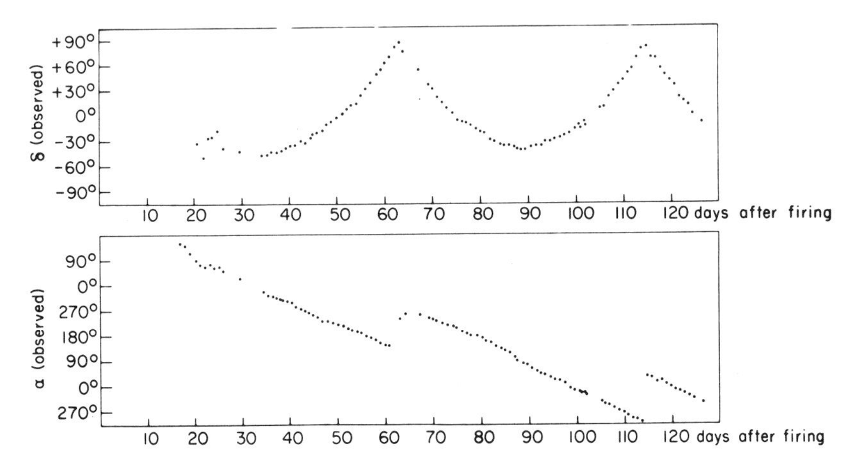

FIG. 2. Right ascension α and declination δ of $\boldsymbol{\Omega}$ vs. time.

of accuracy of the observations, the position of $\boldsymbol{\Omega}$ with respect to the body was observed to be fixed along the axis of maximum moment of inertia (private communication). Finally, a slow decay of the angular momentum $L\boldsymbol{\Omega}$, which is a second-order effect, has been precisely

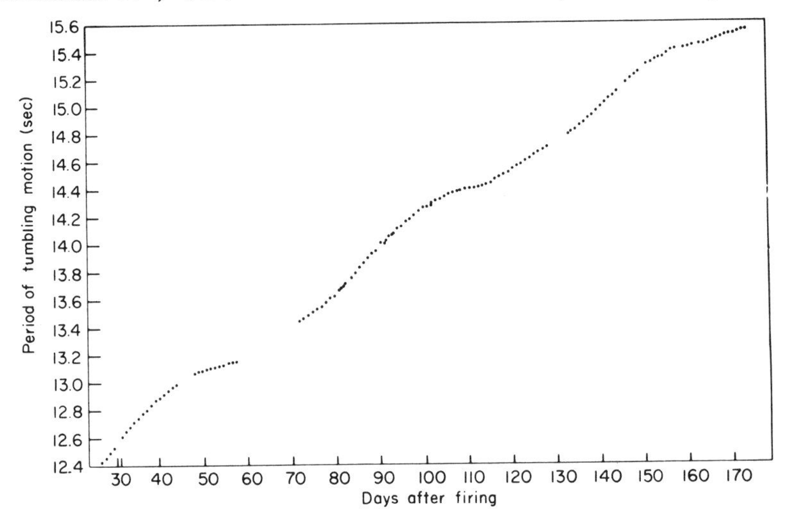

FIG. 3. Period of tumbling motion (seconds) vs. time.

detected and studied, showing interesting features (see Fig. 3). We will distinguish the first-order effects (motion of $\mathbf{\Omega}$) from the second-order effects (slow-down of the tumbling motion).

The precessional motion of $\mathbf{\Omega}$ can be explained by a torque whose components are quadratic functions of the components of $\mathbf{\Omega}$, that is, a torque with the same geometric characteristics as the gravitational torque. However, the magnitude of the latter, which can be very well determined, is between one-half and one-third of the needed amount; and the orientation is quite different.

In this case of a satellite with a very large perigee distance and a complicated shape, but without any large amount of hard magnetic materials in its structure, the aerodynamic, eddy-current, and hysteresis torques are two orders of magnitude less than the gravitational torque [3]. Only the magnetic torque coming from the interaction between the earth's magnetic field and the equivalent magnetic dipole of the satellite appears sufficient to explain the motion of $\mathbf{\Omega}$, as we have already observed in the case of Explorer IV [4].

The torque produced by the earth's magnetic field is

$$\mathbf{M} = \mathbf{I} \times \mathbf{H} \tag{1}$$

where $\mathbf{I}$ is the vector of the equivalent magnetic dipole of the satellite. Neglecting the second-order effects, we can write for the equivalent magnetic dipole of the satellite

$$\mathbf{I} = \mathbf{I}_0 + \mathbf{m}(\mathbf{H}) \tag{2}$$

where $\mathbf{I}_0$ is the intrinsic magnetization vector from the magnetized components and interior current loops; and $\mathbf{m}(\mathbf{H})$ is the magnetic dipole induced by the interaction of the earth's magnetic field with the ferromagnetic components. It seems reasonable to suppose that the permeable components of the satellite have, in the complex, the same property of symmetry with respect to the figure axis Z of the satellite as the exterior shape has. Therefore, neglecting hysteresis effects, we may write Eq. (1) as

$$\mathbf{M} = \mathbf{I}_0 \times \mathbf{H} + \{\mu_1(\mathbf{H} \cdot \mathbf{K})\,\mathbf{K} + \mu_2(\mathbf{H} \cdot \mathbf{K} \times \mathbf{\Omega})\,\mathbf{K} \times \mathbf{\Omega} + \mu_3(\mathbf{H} \cdot \mathbf{\Omega})\,\mathbf{\Omega}\} \times \mathbf{H} \tag{1'}$$

Here μ_1, μ_2, and μ_3 are coefficients related to the geometric shape and the magnetic property of the permeable components [7], and $\mathbf{K}$ is a unit vector along the Z axis.

When we average over a tumbling cycle we will find that

$$\mathbf{M} = (\mathbf{I}_0 \cdot \mathbf{\Omega})\,\mathbf{\Omega} \times \mathbf{H} + \mu^*(\mathbf{H} \cdot \mathbf{\Omega})\,\mathbf{\Omega} \times \mathbf{H} \tag{1''}$$

where $\mu^* = \mu_3 - \frac{1}{2}(\mu_1 + \mu_2)$. It seems to me important to observe that for evaluating μ_1, μ_2, and μ_3, we must take into account that, while μ_1 and μ_2 are related to a component of the earth's magnetic field sinusoidally changing in a tumbling period (13 seconds), μ_3 is related to a very slowly changing component of the same field.

Before the launching no measurements of the intrinsic magnetic dipole of the payload were made; only a crude evaluation of the intrinsic magnetic dipole of the last stage of the rocket was carried out. Therefore, we do not know $\mathbf{I}_0$ with enough accuracy. All we know about the permeable structure of the satellite is that the material of the last stage of the rocket is 410 stainless steel, that small elongated cylinders of permalloy material were put into the payload for shielding purposes, and finally that there is an iron annular plate in the tail, lying in a plane normal to the Z axis. We do not know anything about the permeable structure of the radio-transmitter complex. In order to obtain a possible explanation of the observed precessional motion of $\mathbf{\Omega}$, we must examine the values of I_Ω and μ^* that give the best agreement with the observations. For this purpose we need better observations of the orientation of $\mathbf{\Omega}$ and a precise numerical integration. Since the probable error in the orientation of $\mathbf{\Omega}$ is of the order of a few degrees, we studied separately, using the very quick averaging procedure, the cases where I_Ω or μ^* is negligible. In both we found the possibility of having good agreement with the observed motion of $\mathbf{\Omega}$, if $\mu^* = 4\pi \times 4.76 \times 10^5$ emu, $I_\Omega = 0$ and also if $\mu^* = 0$, $I_\Omega = 0.63$ amp-meter^2. While the needed value of μ^* seems too high, an estimate of the actual value is very difficult. The needed value of I_Ω in the second case is certainly not in good agreement with the measurement made on the last stage of the rocket before launching, although no measurements have been made for the payload. Our opinion is that perhaps the contribution of both magnetic dipoles, the induced and the intrinsic, are significant. In any case, our goal is to give a very easy method for a first-approximation study of the phenomenon. In a second stage this method may be improved by numerical integration on a high-speed computer.

We assume that the following hypotheses are satisfied:

(a) The axis of rotation of $\mathbf{\Omega}$ is almost fixed in the body in the meaning we will state precisely later.

(*b*) The permeable structures of the satellite are such that we can write Eq. (1′) neglecting second-order quantities.

(c) The aerodynamic, hysteresis, and eddy-current torques are negligible for the explanation of the main precessional motion of Ω.

We prefer to leave (a) as an hypothesis since we think a theoretical study of the mechanism of the stabilization of the $\mathbf{\Omega}$ axis in a region very close to the axis of maximum moment of inertia would require a good knowledge of the internal dissipation of energy (nutation damper). The $\mathbf{\Omega}$-axis cannot be precisely fixed in the body: If it were, it would also be fixed with respect to a fixed reference system and we would not have the observed precessional motion. However, the total angular-velocity vector of the satellite may be considered as the sum of the angular velocity vector $\omega\mathbf{\Omega}$ of the tumbling motion, fixed in the body, and the precessional velocity of the vector $\mathbf{\Omega}$, with respect to a fixed reference system. The magnitude of this second vector is of the order of 10^{-5} of the value of ω. This means that the total angular velocity vector may always be very close to the tumbling axis (less than 1 second of arc separation), which is consistent with the observed motion.

In hypothesis (b) we prefer to leave the parameters μ_i undetermined in view of our poor knowledge of the magnetic properties of the permeable components and the fact that these components are moving in the weak magnetic field of the earth. As we have already stated Dr. C. Lundquist made, before launch, a crude evaluation of the magnetization of the last stage of the rocket; his results gave a magnetization vector with a large component in the direction of the Z axis and a small component (1/20 of the former) in a transversal direction that was not very well defined. For hypothesis (c) we can say first that the magnitudes of these torques are of two orders of magnitude less than the gravitational torque, and second that their orientation would not be in good agreement for the explanation of the observed variation in the direction of $\mathbf{\Omega}$. The aerodynamic torque can be fairly precisely determined using the hypothesis of neutral drag; the hysteresis and eddy-current torques cannot be determined so precisely. In any case, however, all these torques are dissipative, and the dissipation process involves torques of the order of 1 dyne-cm and not of 100 dyne-cm; the latter amount would be needed to explain the variation of the orientation of $\mathbf{\Omega}$ without a dissipation of the same order of magnitude.

The validity of the following procedures for the deduction of the equation of motion is postulated:

(d) To compute the torque acting on the satellite, we average over 1 period of rotation of the body around $\mathbf{\Omega}$ (tumbling period).

(e) We average the torque over 1 orbital period of the satellite.

(f) Finally, we average over 1 day.

The procedures followed in (d) and (e) are the usual ones used in the perturbation method for determining the gravitational and aerodynamic torques [5]. In one tumbling period, the center of mass of the satellite will move along a 100-km arc of the orbit. In the case of Explorer XI, this amount corresponds to a 1° variation in the orientation of the radius vector from the earth's center E to the satellite's center of mass G; that means that the variation in the field is 2 orders of magnitude less than the intensity of the field. Furthermore, in 1 orbital period the variation of the orientation of $\mathbf{\Omega}$ is of the order of 0.5°. We prefer to use procedures (d) and (e) as working hypotheses since they also give good results for the gravitational torque. Dr. Leland Cunningham made for the Huntsville Center a step-by-step integration of the original equations to determine the effects of the gravitational torque on Explorer XI. His results were the same as those obtained by the averaging procedure (private communication from Dr. Lundquist). As for (f), we prefer to use the same averaging procedure. Therefore, we are able to arrive quickly at the interesting results that follow.

At the end of this report we will make a preliminary analysis of the slow-down of the spinning motion. This slow-down is definitely a consequence of the eddy-current torque and hysteresis effects, since the aerodynamic torque is very small (less than 0.1 dyne-cm). Both effects are proportional to the square of the component $H_{\perp}$ of $\mathbf{H}$ normal to the tumbling axis. The correlation between the square of this component averaged over 1 orbital period of the satellite and the first derivative of the tumbling period states the nature of the breaking torque, even if it seems more complicated, but not hopeless, to distinguish between the two effects [3].

A more detailed analysis has been made at the Massachusetts, Institute of Technology. This analysis of the observations (a dozen per day) of the variation of the period seems to show definitely a term with the period of 1 day, which we think is correlated with the variation in 1 day of the position of the earth's magnetic dipole, which in 24 hours rotates around the earth's geographic axis.

The averaging procedure we used in our computation is related to the accuracy of the observations of $\mathbf{\Omega}$. In view of the good observations of ω, a more accurate knowledge of the value of this parameter may make a numerical integration of the general equation of motion worthwhile.

II. Evaluation of Torques and Analysis of Motion

A. Magnetic Torque from Induced Magnetic Dipole

Let us compute the effective magnetic torque coming from the induced magnetic dipole. From Eq. (1''), assuming $\mathbf{I}_0 \cdot \mathbf{\Omega} = 0$, we have

$$\tilde{\mathbf{M}} = \mu^*(\mathbf{H} \cdot \mathbf{\Omega})\, \mathbf{H} \times \mathbf{\Omega} \tag{3}$$

In the usual notation, let

$$r\mathbf{r} = \frac{a(1 - e^2)}{1 + e \cos(\theta - \omega)} \{\cos\theta\, \mathbf{i} + \sin\theta \cos i\, \mathbf{j} + \sin\theta \sin i\, \mathbf{k}\} \tag{4}$$

be the vector equation of the motion of the satellite's center of mass G. Here $\mathbf{r}$ is the unit vector of the direction EG from the earth's center; $\mathbf{i}$ is the unit vector of the ascending node in the equatorial plane of the earth's equivalent (magnetic) dipole; and $\mathbf{k}$ is the direction of the axis of this dipole. We assume for the earth's magnetic field the usual first approximation [6].

$$\mathbf{H} = \frac{\mu_E}{r^3}(\mathbf{k} - 3(\mathbf{k} \cdot \mathbf{r})\, \mathbf{r}) \tag{5}$$

The average value in one orbital period is given by the integral

$$\tilde{\mathbf{M}} = \frac{\mu^*}{T} \int_0^T (\mathbf{H} \cdot \mathbf{\Omega})\, \mathbf{\Omega} \times \mathbf{H}\, dt \tag{6}$$

Since

$$\frac{d\theta}{dt} = \frac{2\pi a^2(1 - e^2)^{1/2}}{Tr^2} \tag{7}$$

we obtain from Eq. (3), (4), (5), and (6) the following expression:

$$\tilde{\mathbf{M}} = \frac{\mu^*}{2\pi a^2(1 - e^2)^{1/2}} \mathbf{\Omega} \times \int_0^{2\pi} r^2(\mathbf{H} \cdot \mathbf{\Omega})\, \mathbf{H}\, d\theta \tag{8}$$

From Eqs. (3) and (4) we have

$$\mathbf{H} = \mu_E \frac{[1 + e\cos(\theta - \omega)]^3}{a^3(1 - e^2)^3} \{-3 \sin\theta \cos\theta \sin i\, \mathbf{i} - 3 \sin^2\theta \sin i \cos i\, \mathbf{j} + (1 - 3\sin^2\theta \sin^2 i)\, \mathbf{k}\} \tag{9}$$

Let Ω_1, Ω_2, Ω_3 be the components of $\mathbf{\Omega}$ with respect to $(\mathbf{i}, \mathbf{j}, \mathbf{k})$; then from Eq. (7), neglecting some small terms in e^2, we finally obtain

$$\tilde{\mathbf{M}} = \frac{\mu^* \mu_E{}^2}{a^6(1 - e^2)^{9/2}} \mathbf{\Omega} \times \left\{ \frac{9}{8} \Omega_1 \mathbf{i} \sin^2 i + \frac{27}{8} \Omega_2 \mathbf{j} \sin^2 i \cos^2 i \right.$$
$$+ \left(1 - 3 \sin^2 i + \frac{27}{8} \sin^4 i\right) \Omega_3 \mathbf{k} \tag{10}$$
$$\left. - \frac{3}{2} \sin i \cos i \left(1 - \frac{9}{4} \sin^2 i\right) (\Omega_2 \mathbf{k} + \Omega_3 \mathbf{j}) \right\}$$

B. Gravitational Torque

We shall now compute the gravitational torque acting on the satellite. Let $(\mathbf{N}, \mathbf{W}, \mathbf{U})$ be an orthogonal reference frame centered at E, where $\mathbf{N}$ is the unit vector of the orbit's ascending node in the geographic equator, and $\mathbf{U}$ is the unit vector of the earth's axis. Also, let i_0 be the inclination of the orbit with respect to the geographic equator, and $\mathbf{n}$ be the unit vector normal to the orbital plane. Averaging the gravitational torque $\mathbf{G}$ over one tumbling period and then over one orbital revolution of G, we have, finally,

$$\tilde{\mathbf{G}} = \tfrac{3}{4} \tilde{\omega}^2 (A - C)\, (\mathbf{\Omega} \cdot \mathbf{n})\, \mathbf{\Omega} \times \mathbf{n} \tag{11}$$

Here C is the moment of inertia of the satellite with respect to the $\mathbf{\Omega}$ axis, and A is the moment of inertia with respect to an axis normal to $\mathbf{\Omega}$ and

$$\tilde{\omega}^2 = \frac{h}{T} \int_0^T \frac{dt}{r^3} \tag{12}$$

where h is the characteristic constant of the earth's gravitational attraction. For Explorer XI, we have

$$\begin{aligned} i_0 &= 28.8 \text{ deg} \\ \mathbf{n} &= -\sin i_0 \mathbf{W} + \cos i_0 \mathbf{U} \end{aligned} \tag{13}$$

and

$$\beta = \tfrac{3}{4} \tilde{\omega}^2 (A - C) = 1.2 \times 10^2 \text{ dyne-cm.} \tag{14}$$

The components of $\tilde{\mathbf{G}}$ with respect to (**N**, **W**, **U**) are the following:

$$G_x = \frac{1}{2}\beta\Omega_y\Omega_z + \frac{\sqrt{3}}{4}\beta(\Omega_z^2 - \Omega_y^2)$$

$$G_y = -\frac{3}{4}\beta\Omega_x\Omega_z + \frac{\sqrt{3}}{4}\beta\Omega_x\Omega_y \tag{15}$$

$$G_z = \frac{1}{4}\beta\Omega_x\Omega_y - \frac{\sqrt{3}}{4}\beta\Omega_x\Omega_z$$

where Ω_x, Ω_y, Ω_z are the components of $\mathbf{\Omega}$ with respect to (**N**, **W**, **U**).

C. Differential Equation of Perturbed Motion

The equation of motion

$$\frac{L\,d\mathbf{\Omega}}{dt} = \tilde{\mathbf{M}} + \tilde{\mathbf{G}} \tag{16}$$

is now to be projected onto an inertial reference frame and then integrated with initial conditions corresponding to an observed orientation at the chosen initial time. We prefer to project Eq. (16) onto the moving reference system (**N**, **W**, **U**). We need the equations for passing from the reference system (**i**, **j**, **k**) to the reference system (**N**, **W**, **U**). Let I be the angle $\widehat{\mathbf{kU}}$; ψ, the angle **N** makes with the intersection of the geomagnetic equator and the geographic equator; and α, the angle $\widehat{\mathbf{iN}}$. We have first

$$\begin{aligned} \tan\alpha &= -\frac{2\tan(\psi/2)\sin I}{\sin(i_0 - I) + \tan^2(\psi/2)\sin(I_0 + i_0)} \\ \cos i &= \cos i_0 \cos I + \sin i_0 \sin I \cos\psi \end{aligned} \tag{17}$$

and also

$$\begin{aligned} \mathbf{i} &= \cos\alpha\,\mathbf{N} + \sin\alpha\cos i_0\mathbf{W} + \sin\alpha\sin i_0\mathbf{U} \\ \mathbf{j} &= (-\sin I\sin i_0\sin\alpha\cos\psi - \cos I\sin i_0\sin\alpha)\,\mathbf{N} \\ &\quad + (\cos I\cos\alpha - \sin\alpha\sin i_0\sin I\sin\psi)\,\mathbf{W} \\ &\quad + \sin I(\sin\psi\sin\alpha\cos i_0 + \cos\psi\cos\alpha)\,\mathbf{U} \\ \mathbf{k} &= \sin I\sin\psi\,\mathbf{N} - \sin I\cos\psi\mathbf{W} + \cos I\mathbf{U} \end{aligned} \tag{18}$$

Here $I = 11.5$ deg; and $\psi = 2\pi t$ rad/day. In our approximation we will obtain from Eq. (17) and (18)

$$\begin{aligned} \sin\alpha &= -0.41\sin\psi, \qquad \cos\alpha = 1 - 0.08\cos\psi \\ \sin i &= 0.5 - 0.18\cos\psi, \qquad \cos i = 0.86 + 0.1\cos\psi \end{aligned} \tag{17'}$$

and

$$\begin{aligned} \mathbf{i} &= (1 - 0.08\cos\psi)\,\mathbf{N} - 0.36\sin\psi\mathbf{W} - 0.2\sin\psi\,\mathbf{U}, \\ \mathbf{j} &= (0.35\sin\psi + 0.02\sin 2\psi)\,\mathbf{N} + (1 - 0.08\cos\psi \\ &\quad -0.02\cos 2\psi)\,\mathbf{W} + (-0.04 + 0.2\cos\psi)\,\mathbf{U} \\ \mathbf{k} &= 0.2\sin\psi\mathbf{N} - 0.2\cos\psi\mathbf{W} + 0.98\,\mathbf{U}. \end{aligned} \tag{18'}$$

It is necessary now to note that for Explorer XI the reference system (**N**, **W**, **U**) is rotating around the **U** axis in a uniform motion with an angular velocity (regression of the node) of -0.087 rad/day. This means that Eq. (11) projected over the chosen reference system takes the form

$$\begin{aligned} L\left(\frac{d\Omega_x}{dt} + 0.087\Omega_y\right) &= \mathrm{M}_x(\Omega_x, \Omega_y, \Omega_z, t) + G_x \\ L\left(\frac{d\Omega_y}{dt} - 0.087\Omega_y\right) &= M_y(\Omega_x, \Omega_y, \Omega_z, t) + G_y \\ L\,\frac{d\Omega_z}{dt} &= M_z(\Omega_x, \Omega_y, \Omega_z, t) + G_z \end{aligned} \tag{19}$$

The functions G_x, G_y, G_z are quadratic functions of Ω_x, Ω_y, Ω_z with constant coefficients [Eq. (13)]; M_x, M_y, M_z are also quadratic functions of Ω_x, Ω_y, Ω_z but the coefficients are periodic functions of t through ψ, with a period of 1 day. It would be possible to perform a numerical integration corresponding to some initial condition similar to the observed conditions of the 35th day after firing. We choose this day since after this we have good observations of the precessional motion of $\mathbf{\Omega}$. The amount of work involved in this numerical computation, even if worthwhile, suggested to us that we use a first approximation to compute the average values of the components M_x, M_y, M_z over 1 day.

The observed numerical variation of the orientation of $\mathbf{\Omega}$ is not more than 10 deg per day: in averaging, we consider the orientation of $\mathbf{\Omega}$ constant, using the mean orientation for the day. The displacement of $\mathbf{\Omega}$ from the mean value for the day is not greater than 5 deg. It is difficult to evaluate the error involved in this averaging procedure, but in any case we think that the approximation is quite good. We prefer to follow this method to confirm quickly our feelings about the nature of the torque needed to explain the precessional motion.

D. Evaluation of the Induced Dipole

We shall now discuss and integrate the differential system obtained by the procedure explained above. Let us put

$$\gamma = \frac{\mu^* \mu_E{}^2}{a^6(1 - e^2)^{9/2}} \tag{20}$$

For our case, a quick evaluation gives

$$\gamma = 1.4 \times 4\pi\mu^* \times 10^{-3} \text{ dyne-cm} \tag{21}$$

Averaging over 1 day we obtain

$$\mathbf{M} = \gamma\mathbf{\Omega} \times \{0.32\Omega_x\mathbf{N} + 0.62\Omega_y\mathbf{W} + 0.47\Omega_z\mathbf{U} - 0.25(\Omega_z\mathbf{W} + \Omega_y\mathbf{U})\} \tag{22}$$

It follows that

$$\begin{aligned} M_x &= -0.15\gamma\Omega_y\Omega_z + 0.25\gamma(\Omega_z{}^2 - \Omega_y{}^2) \\ M_y &= -0.15\gamma\Omega_x\Omega_z + 0.25\gamma\Omega_y\Omega_x \\ M_z &= 0.30\gamma\Omega_x\Omega_y - 0.25\gamma\Omega_x\Omega_z \end{aligned} \tag{23}$$

To write Eq. (16) in the explicit form, we have to evaluate $L = A\omega$, where ω is the observed angular velocity. We find that

$$L = 1.616 \frac{2\pi}{13} \cdot 10^8 \text{ gm-cm}^2\text{-sec} \tag{24}$$

Choosing the day as the unit of time, now we can write, finally, Eq. (16) and we obtain

$$\begin{aligned} 910 \frac{d\Omega_x}{dt} &= -(0.15\gamma + 60)\, \Omega_y\Omega_z + (0.25\gamma + 52)(\Omega_z{}^2 - \Omega_y{}^2) - 79\Omega_y \\ 910 \frac{d\Omega_y}{dt} &= -(0.15\gamma + 90)\, \Omega_x\Omega_z + (0.25\gamma + 52)\, \Omega_y\Omega_x + 79\Omega_x \\ 910 \frac{d\Omega_z}{dt} &= (0.30\gamma + 30)\, \Omega_x\Omega_y - (0.25\gamma + 52)\, \Omega_x\Omega_z \end{aligned} \tag{25}$$

The differential system in Eq. (25) has two first integrals: the obvious one,

$$\Omega_x^2 + \Omega_y^2 + \Omega_z^2 = 1 \tag{26}$$

and

$$\tfrac{1}{2}(0.15\gamma + 90)\,\Omega_z^2 + \tfrac{1}{2}(0.30\gamma + 30)\,\Omega_y^2 + (0.25\gamma + 52)\,\Omega_y\Omega_z - 79\Omega_z = E \tag{27}$$

The intersection of the sphere in Eq. (26) with the cylinder in Eq. (27) gives the path of the vertex of the vector $\mathbf{\Omega}$ with respect to the rotating reference system (**N**, **W**, **U**).

In Fig. 4(a) and 4(b) we plotted the projections on the **WU** plane and the **NU** plane of the observed position of the vertex of $\mathbf{\Omega}$. If we have

$$\mu^* = 4\pi \times 4.76 \times 10^5 \text{ emu} \tag{28}$$

we will obtain

$$\gamma = 6.6 \times 10^2 \tag{29}$$

Consequently Eq. (27) becomes

$$95\Omega_z^2 + 115\Omega_y^2 - 220\Omega_y\Omega_z - 79\Omega_z = E \tag{30}$$

Equation (30) represents a family of hyperbolas. If we choose $E = 49.5$, we will obtain the hyperbola, shown in Fig. 4(a). Figure 4(b) represents the projection on the $\Omega_x\Omega_z$ plane of the path Γ of $\mathbf{\Omega}$ corresponding to the arc AB of the hyperbola. The motion of $\mathbf{\Omega}$ with respect to the reference system (**N**, **W**, **U**) is periodic. The good agreement of the observed path emerges clearly [see Fig. 4(c)].

We also need to compare the equation of motion along the path with the observations. The easiest way to do this is to compare the observed values of $d\Omega_z/dt$ as functions of Ω_z, with the value of the same derivative computed from the third part of Eq. (25), which, when we take into account Eq. (29), becomes

$$\frac{d\Omega_z}{dt} = 0.25\Omega_x\Omega_y - 0.24\Omega_x\Omega_z \tag{31}$$

In Fig. 4(c) we plotted the values of the second term of Eq. (31) as a function of Ω_z, evaluated using Eq. (26) and (27) with $E = 49.5$. The *dots* are the observed values of the same derivative $d\Omega_z/dt$ as a function Ω_z. The good agreement is evident.

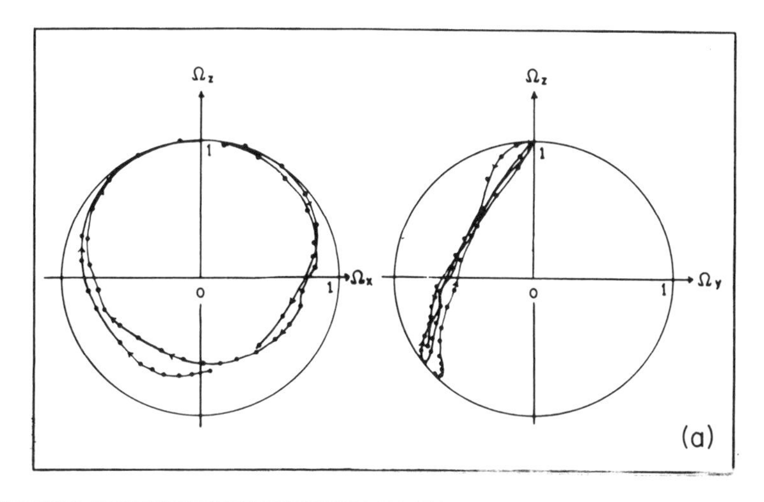

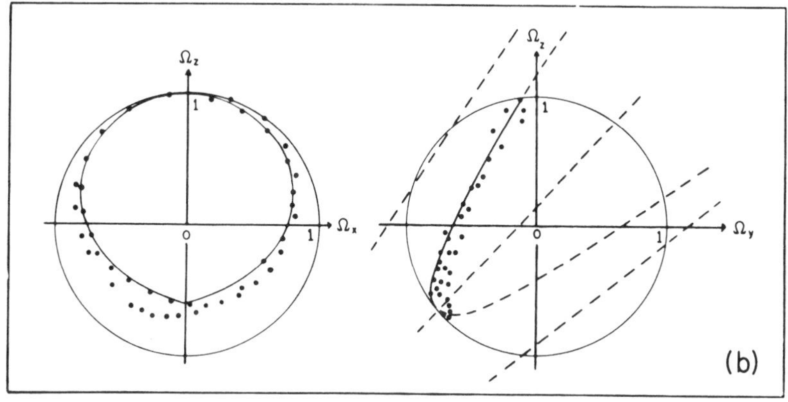

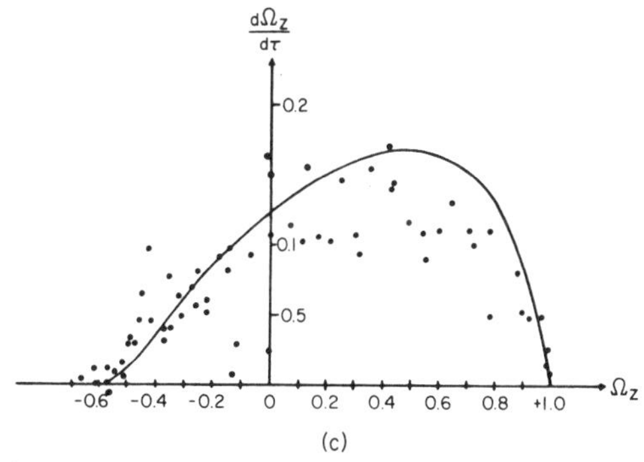

FIG. 4. (a) and (b) Projection of the observed path of the vertex of $\mathbf{\Omega}$ over the xz and yz planes of a rotating reference system (x, ascending node; z, earth's axis). Solid lines are computed path and dots are observed position. (c) Computed (solid lines) and observed (dots) value of $d\Omega_z/dt$ vs. Ω_z.

E. Evaluation of the Intrinsic Dipole

We shall now show that a component of the permanent magnetization (evaluated as approximately 0.630 amp-meter²) normal to the Z axis can explain the observed motion of $\mathbf{\Omega}$.

Using the same procedure as above, we will find [4]

$$L\frac{d\mathbf{\Omega}}{dt} = \lambda\mathbf{\Omega} \times \left(\frac{5}{8}\mathbf{U} - \frac{3\sqrt{3}}{8}\mathbf{W}\right) \tag{32}$$

where

$$\lambda = \frac{\mu_E \mu^*}{a^3(1-e^2)^{3/2}} \tag{33}$$

If $\mu^* = -0.630$ amp-meter², which corresponds to a value of λ of 126 dyne-cm, we will have for a first integral

$$45\Omega_y^2 + 15\Omega_y^2 - 52\Omega_y\Omega_z - 82\Omega_y = E \tag{34}$$

Therefore, Eq. (30) and (31) become

$$910\frac{d\Omega_z}{dt} = 30\Omega_x\Omega_y - 82\Omega_x - 52\Omega_x\Omega_z \tag{35}$$

and

$$45\Omega_z^2 + 15\Omega_y^2 - 52\Omega_y\Omega_z - 82\Omega_y = E \tag{36}$$

We have again a family of hyperbolas. We choose the constant E in such a way that the curve given by Eq. (36) passes through the point $\Omega_y = -0.8$, $\Omega_z = -0.6$; that is, $E = 66.5$.

In Fig. 5(a) we have the projection of the path of $\mathbf{\Omega}$ on the $\Omega_x\Omega_z$,

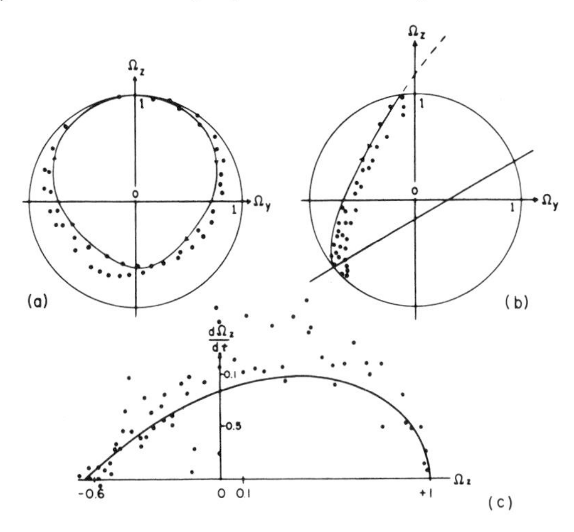

Fig. 5. Same as Fig. 4(c), corresponding to the hypothesis of a constant component of the permanent magnetization in the direction of $\mathbf{\Omega}$.

$\Omega_y\Omega_z$ planes, corresponding to $\mu^* = 0.630$ amp-meter². The same results would be obtained if we assumed the satellite had a residual spin motion and a large axial component of the permanent magnetization.

While a residual spin velocity of the needed amount was not observed (private communication from Prof. Kraushaar, Massachusetts, Institute of Technology), the existence of the needed component of the residual magnetization has to be postulated.

F. Period of Tumbling

In order to explain the variation in the period of the tumbling motion we computed the value of the square of the component $H_\perp$ of $\mathbf{H}$ normal to $\mathbf{\Omega}$. For a first approximation, we averaged over 1 day, and then plotted the value $\tilde{H}_\perp^2/\tilde{H}^2$ *versus* time and the value of the first derivative of the period in seconds day in Fig. 6. The correlation looks good, at least for the positions of the maxima and minima. We have to take into account both the averaging procedure and the error of the observations of $\mathbf{\Omega}$, which also affects the computed value of $\tilde{H}_\perp^2$.

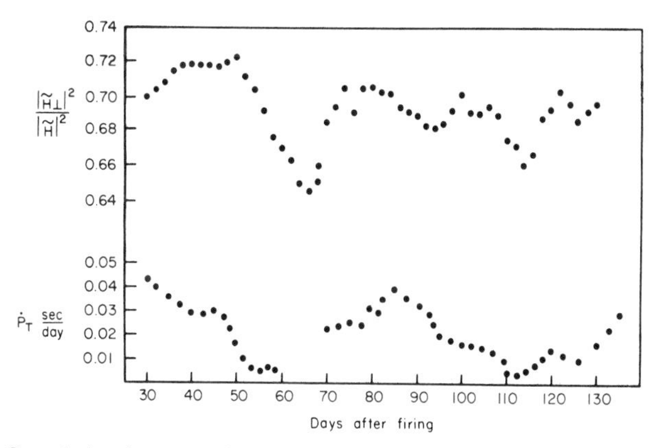

FIG. 6. Correlation between the first derivative of the period and the square of the component of $\tilde{\mathbf{H}}$ normal to $\mathbf{\Omega}$.

If we assume the damping torque is the sum of the eddy-current torque (which is assumed proportional to the angular velocity ω) and the hysteresis torque (which is considered independent of ω), we can write

$$M_D = (\sigma\omega + \nu) H_\perp^2 \tag{37}$$

The magnitude of the torque needed to explain the braking is of the order of 1 dyne-cm. The evaluation of the coefficients σ and ν is extremely difficult and requires good information about the physical properties of the conducting and ferromagnetic components of the satellite; we therefore plotted only the first derivative of the tumbling period *versus* time. Since the tumbling period varies from 12.4 to 14.6 seconds during the one-hundred day interval of observations of the orientation of $\mathbf{\Omega}$, the behavior of $d\omega/dt$ cannot be very different, and the position in time of the maxima and minima definitely cannot undergo appreciable changes.

A deeper and more detailed analysis of the observed variation of ω, even during 1 day, is strongly suggested by the accuracy of the observations and by the interesting correlation of the diurnal periodic term with the variation of the position of the earth's magnetic dipole with respect to $\mathbf{\Omega}$. If we take into account that the earth's magnetic dipole makes an angle of 11.5 deg with the earth's geographic axis, the maximum displacement of the earth's magnetic dipole will be of 23 deg in 12 hours, which is much larger than the maximum displacement of Ω (about 5 deg) in the same period. This fact makes the effect detectable.

Acknowledgments

I am indebted to Dr. C. Lundquist and R. Naumann of the Huntsville Center and to Prof. W. L. Kraushaar and Prof. G. Clark of the Massachusetts Institute of Technology for the observational material and for fruitful discussion. I am grateful to Miss Cara Munford for her help in the computation, and to Miss Joan Weingarten for her help in the writing of the paper.

References

1. R. J. Naumann, Recent information gained from satellite orientation measurement. *In* "Ballistic Missiles and Space Technology," Vol. 3, pp. 445-453. Pergamon, New York, 1961.
2. R. J. Naumann, S. Fields, and R. Holland, "Determination of Angular Momentum Vector for the S-15 Payload (Explorer XI). George C. Marshall Space Flight Center, Huntsville, Alabama, 1961.
3. R. H. Wilson, Jr., "Rotational Decay of Satellite 1960 η2 Due to the Magnetic Field of the Earth." NASA Goddard Space Flight Center, Greenbelt, Maryland, 1961.
4. G. Colombo, The Motion of Satellite 1958 Epsilon around Its Center of Mass, *Smithsonian Astrophys. Obs., Spec. Rept. No. 70*, 1961.
5. V. V. Beletsky, Motion of an artificial earth satellite about its center of mass. *In* "Artificial Earth Satellites (L. V. Kurnosova, ed.) Vol. 1, pp. 30-54. Plenum Press, New York, 1960.
6. E. J. Chernosky, and E. Maple, Geomagnetism. *In* "Handbook of Geophysics," Chapter 10, pp. 10-11 to 10-68. MacMillan, New York, 1960.
7. L. Spitzer, Space telescopes and components. *Astronaut. J.* **65**, 242-263 (1960).

An Investigation of the Observed Torques Acting on Explorer XI

ROBERT J. NAUMANN

Research Projects Division, George C. Marshall Space Flight Center, National Aeronautics and Space Administration, Huntsville, Alabama

I. INTRODUCTION

SHORTLY AFTER THE LAUNCHING of Explorer IV in July 1958 there arose an active interest in satellite orientation analysis. The reason for this interest was twofold: some unusual fluctuations in interior temperature were observed which apparently could be explained only by a gradual change in the orientation of the satellite angular momentum vector [1]; and the satellite contained a directional radiation detector whose output exhibited definite modulations due to its motion about the satellite center of mass relative to the magnetic field [2]. Since in those early days of space exploration there were severe limitations in payload, no on-board device was included for attitude sensing. Hence, it became necessary to use ground observations of the satellite radio signal strength patterns to deduce the spatial orientation of the satellite [3]. The antenna and recording techniques used for Explorer IV were far from optimum for this type of analysis; nevertheless, several interesting findings resulted. It was definitely established that the angular momentum vector did indeed change its spatial orientation by as much as 10–15 deg per day. Furthermore, the observed change in orientation affected the insolated area in such a manner as to account for the observed temperature fluctuations. The orientation analysis also resulted in the determination of the directional intensity and mirror point distribution of the trapped particles in the inner Van Allen belt [4, 5].

Manuscript received March 1962.

While it was clearly established that the spatial orientation of the angular momentum vector underwent a definite change, the external torques responsible for this change have not been fully understood.[1]

The Explorer XI gamma ray telescope satellite was dynamically similar to Explorer IV. Therefore we were quite interested in performing the same type of analysis on this satellite to see if similar orientation changes would occur. Also, such a radio signal strength analysis would serve as a useful back-up for the optical horizon sensing devices included on board for determining the telescope orientation. Our experience with Explorer IV suggested many improvements in the tracking antenna, data recording, and analysis [6]. The resulting determination appears to be quite accurate with a probable error of generally less than 3 deg, and, as was the case with Explorer IV, a precession of the angular momentum vector at the rate of approximately 10 deg per day was observed.

II. Method of Orientation Determination

Explorer XI has a configuration such that the axis of symmetry is a principle axis of least moment of inertia, as may be seen in Fig. 1. Originally the satellite was set spinning about this axis before injection into orbit. A slight force-free precession was present due to a slight malalignment between the angular velocity and angular momentum vectors. Such a precession causes energy dissipation in the mercury

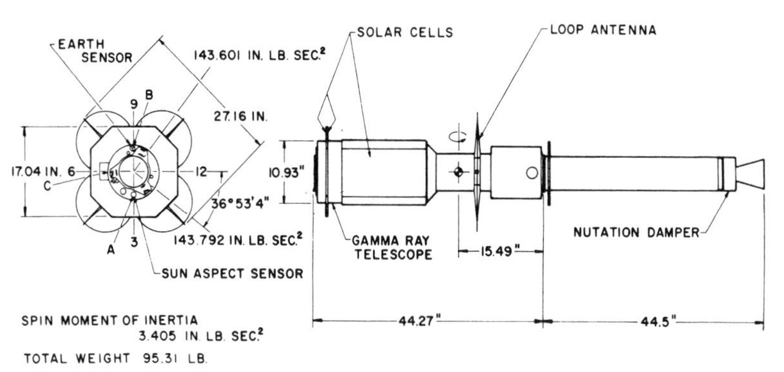

Fig. 1. Explorer XI configuration and dynamic properties.

[1] More recent work indicates that permanent magnetic moments were responsible for the observed motions of Explorers IV and XI. A detailed treatment is given by R. J. Naumann, Observed torque-producing forces acting on satellites, *in "Dynamics of Satellites"* (I.U.T.A.M. Symp., Paris, 1962), (M. Roy, ed.), Springer-Verlag, Berlin, 1963.

damper near the nozzle which, in turn, causes the rotation to be about an axis of greater moment of inertia so that angular momentum is conserved while rotational energy is dissipated. Eventually the precession angle approaches $\pi/2$, the rotation is about an axis of maximum moment of inertia, and a stable or minimum energy configuration is attained. The latter configuration will be assumed for the remainder of the discussion. The symmetry axis rotates in a plane referred to as the tumble plane; the angular momentum is directed along the normal to this plane and will be referred to as the tumble axis.

Actually, the satellite was designed such that one of the principle transverse axes has a slightly greater moment of inertia than the other. The mercury damper succeeded in removing all the residual longitudinal angular momentum so that the entire angular momentum is manifested in rotation about the axis of maximum moment of inertia.

The determination of the orientation of the tumble axis was accomplished by successive measurements of the look angle, that is, the angle the observer-to-satellite vector makes with the tumble axis. This measurement is made from the observation of the recorded signal strength pattern during a tumble cycle.

The satellite shell was used to form an asymmetric dipole antenna resulting in the radiation pattern shown in Fig. 2 and 3. The radiation is plane polarized along the symmetry axis, but the receiving antenna consists of an array of helices which is not affected by polarization. Note in Fig. 2 that the forward lobes are lower by about 6 db than the major lobes to the rear. Also, the side nulls occur at 74 deg from the nose.

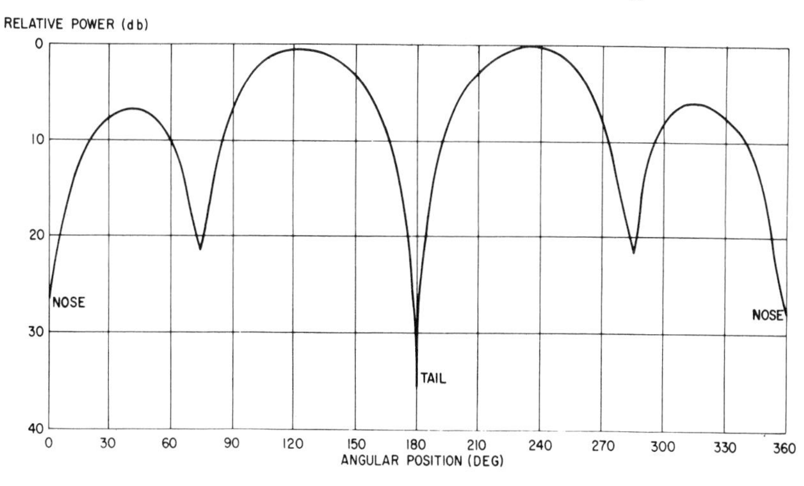

Fig. 2. db Plot of Explorer XI antenna radiation pattern for tracking transmitter (108.06 mc).

The most significant fact observable in the recorded signal strength pattern is the duration of the nose portion, from side null through both nose lobes to the next side null, compared with that of the total tumble period. The ratio of the nose portion to the total period depends on the look angle ψ; hence, ψ can be inferred from a measurement of this ratio.

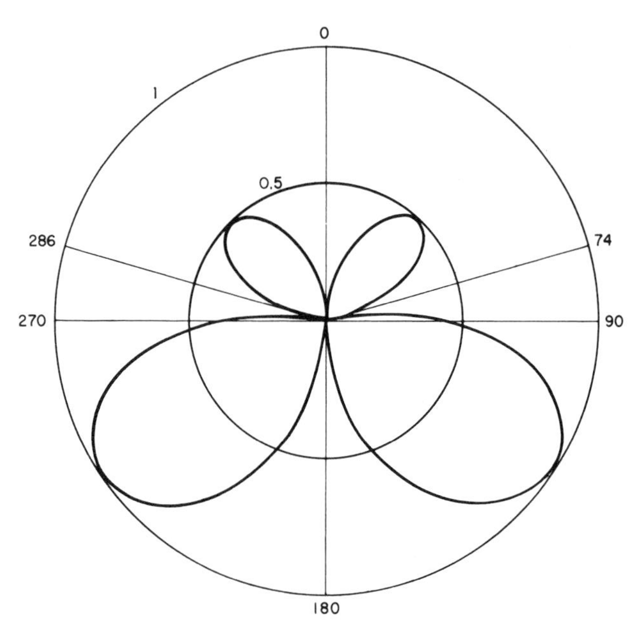

FIG. 3. Polar plot of Explorer XI antenna radiation pattern for tracking transmitter (108.06 mc).

For a look angle of 90 deg, the nose portion is visible for 74 deg out of a half-revolution; hence, the ratio has a maximum value of 0.411. For look angles less than 16 deg, the side lobe is never seen, and hence the ratio is 0. Straightforward geometry results in a plot of this ratio for various look angles shown in Fig. 4. Note the insensitivity for ψ near 90 deg. Because of this, it is necessary to limit this technique to angles somewhat less than 90 deg where the sensitivity is good. Points at which $\psi = 90$ deg may be recognized with good precision by the observation of four very deep nulls during a tumble cycle. This observation is unique for $\psi = 90$ deg, as can be seen from the antenna pattern (Fig. 2). Typical observed signal strength patterns are illustrated in Fig. 5.

Following this procedure, look angles were measured at various times during each pass over the Huntsville station. Since the orbit was such that five successive passes were usually above Huntsville radio horizon,

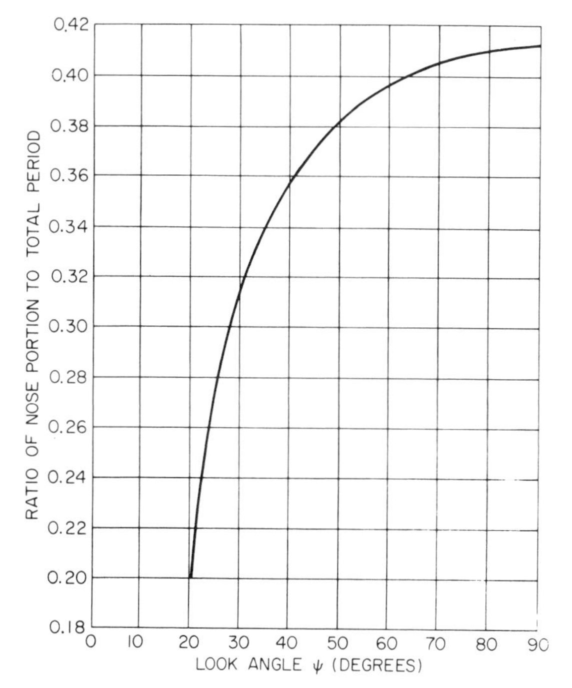

FIG. 4. Plot for obtaining look angle from observed ratio of nose portion to tumble period.

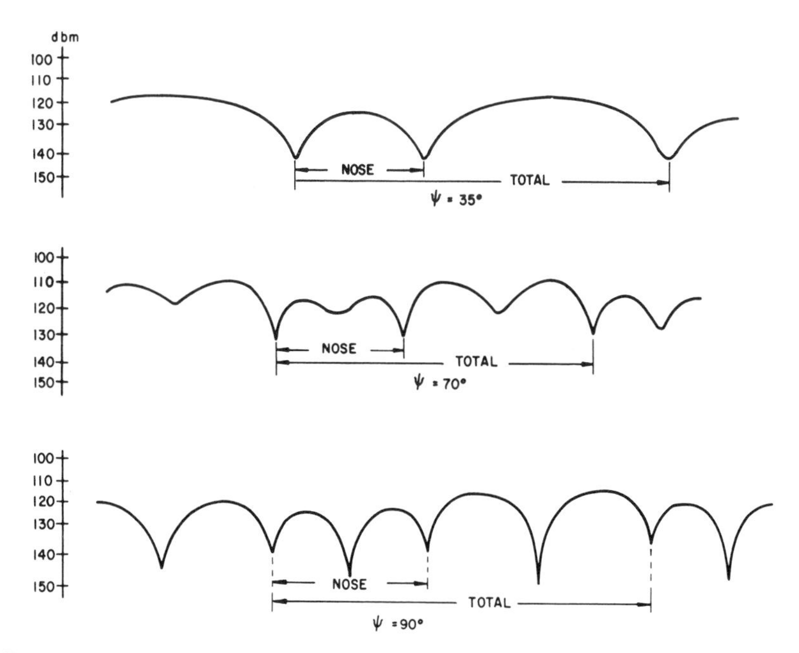

FIG. 5. Typical observed signal strength patterns for various look angles.

the orientation was determined for the time of the middle pass by using look angles measured during the entire set of five passes. This is justified since the time elapsed during the five passes is approximately 8 hours and, for a net motion of 10 deg per day, the tumble axis would move about 3 deg during this interval. This is just the over-all accuracy of the method, and thus there is little to be gained by attempting a pass-by-pass determination. Also, a single pass generally does not have a favorable geometry to define the tumble axis orientation adequately.

The look angles $\psi_{1,2,\ldots,n}$ having been obtained, the tumble axis unit vector must satisfy the following relations:

$$\begin{aligned} \mathbf{U}_1 \cdot \mathbf{L} &= \cos \psi_1 \\ \mathbf{U}_2 \cdot \mathbf{L} &= \cos \psi_2 \\ &\vdots \\ \mathbf{U}_n \cdot \mathbf{L} &= \cos \psi_n \\ |\mathbf{L}| &= 1 \end{aligned} \tag{1}$$

where $\mathbf{U}_{1,2,\ldots,n}$ are the unit vectors from the station to the satellite in a space orientation–fixed coordinate system. These are readily computed from the satellite ephemeris. The constraint equation $|\mathbf{L}| = 1$ is used to eliminate one of the components of **L**. The remaining two components of **L** are computed from the overdetermined set of n equations. This set is made nonlinear by the elimination of the one component so that an iterative scheme is required to find the solution for the two components of **L** that provide the best least square fit. Note that the residuals in $\cos \psi$ are minimized, which tends to introduce a weighting of small values of ψ. This is justifiable since the small values of ψ are inherently more accurate as may be seen in Fig. 4.

There is a quadrant ambiguity in ψ since the observations cannot distinguish quadrant. This is resolved by testing each value of $\mathbf{U}_i \cdot \mathbf{L}$ where **L** is the trial value. If this quantity is positive, $\cos \psi$ is taken as positive, and *vice versa*.

Also there is still the question as to whether the angular momentum is parallel or antiparallel to this **L**. This can be resolved by observing the direction of motion from the on-board horizon sensors. This need only be done once since the excursion of **L** may be followed daily and the day-to-day change is small compared to 180 deg.

The probable error is obtained by comparing each $\cos^{-1} (\mathbf{U}_i \cdot \mathbf{L})$ with the corresponding ψ_i where **L** is the best least square solution. This probable error refers to the accuracy of the look angles rather than the actual components of **L**. This probable error was found to be generally less than 3 deg.

III. Results of the Orientation Determination

The right ascension and declination of the tumble axis resulting from this determination are plotted as a function of time in Fig. 6. It may be seen that the right ascension appears to decrease continuously except when the declination becomes close to 90 deg. At these points,

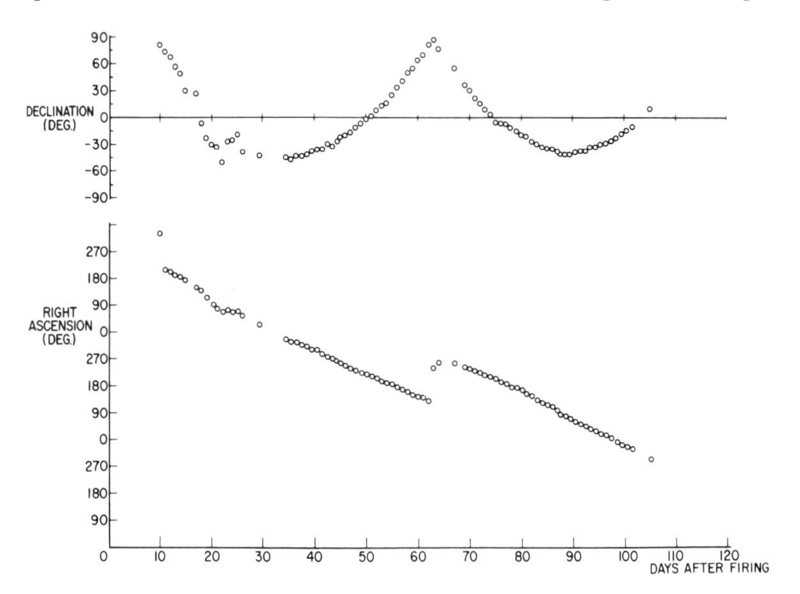

Fig. 6. Right ascension and declination plot of the tumble axis vs. time.

there is an apparent discontinuity in the right ascension which is due to the fact that the right ascension is ill-defined for declinations near 90 deg. On the other hand, the declination appears to oscillate between 90 deg and −43 deg. The torques were determined by fitting polynomials to the right ascension and declination plots in segments, expressing the x, y, z components of the angular momentum in terms of the polynomials, and differentiating with respect to time. Figure 7 shows that these torques are well behaved despite the apparent discontinuity in the right ascension and declination plots. The magnitude of the torques averages 110 dyne-cm with a maximum value of 145 dyne-cm.

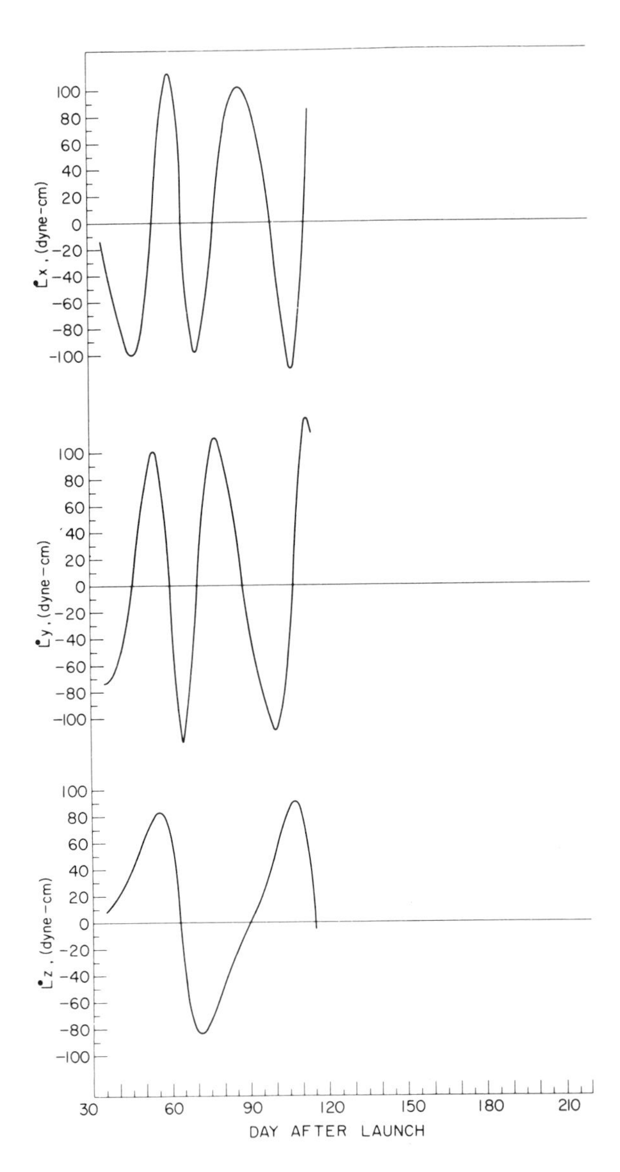

FIG. 7. Components of observed torque in a space orientation—fixed vernal equinox system.

Order-of-Magnitude Considerations of Possible External Torques

1. Gravitational-Gradient Torques

It has been shown by Roberson [7], Doolin [8], and others that the gravitational torques may be found by expanding the potential energy of a body in a Taylor's series about the center of mass, and differentiating with respect to generalized angle coordinates.

For a satellite having cylindrical symmetry about the I_3 axis the resulting torque is:

$$\mathbf{L} = 3\frac{mk^2}{R^3}(I_1 - I_3)(\mathbf{A} \cdot \mathbf{R})(\mathbf{A} \times \mathbf{R}) \tag{2}$$

where **A** and **R** are unit vectors along the satellite symmetry axis and the radius vector, respectively; $(I_1 - I_3)$ is the difference between the transverse and longitudinal moment of inertia; mk^2 is the gravitational constant; and R is the magnitude of the distance from the earth center to the satellite.

For a tumbling satellite, such as Explorer XI, the torque may be averaged over a tumble cycle resulting in

$$\mathbf{L} = \frac{3}{2}\frac{mk^2}{R^3}(I_1 - I_3)(\mathbf{L} \cdot \mathbf{R})(\mathbf{R} \times \mathbf{L}) \tag{3}$$

where **L** is a unit vector along the tumble axis.

For Explorer XI, $(I_1 - I_3)$ is 1.587×10^8 gm cm^2, and the semimajor axis or average R is 7512 km. If we take mk^2 to be 3.986×10^5 km^3/sec^2 and the geometry factor $(\mathbf{L} \cdot \mathbf{R})(\mathbf{L} \times \mathbf{R})$ to have a maximum value of 0.5, the gravitational torque is 113 dyne-cm. Averaging this over an orbital revolution reduces the effective torque by a factor of 2, resulting in 57 dyne-cm. While this torque is significant, it cannot fully explain the observed precession of the tumble axis.

2. Aerodynamic Torques

The aerodynamic torque is maximum at perigee when the satellite is broadside to the velocity vector. Assuming free molecular flow with a drag coefficient $C_D = 2.2$, an area $A = 4200$ cm^2, the velocity $V = 8 \times 10^5$ cm/sec, and the atmospheric density $\rho = 3.8 \times 10^{-15}$ gm/cm^3, the drag force is

$$F_D = \frac{C_D}{2} A\rho V^2 = 11.2 \text{ dynes} \tag{4}$$

The displacement of the center of area from the center of mass is approximately 15 cm. Therefore, the maximum aerodynamic torque is 168 dyne-cm. However, the tumbling motion results in an equal and opposite torque a half tumble cycle later; hence, the first-order torques average to zero owing to the rotational symmetry. Some second-order torques may arise from the difference in the diffuse reflection coefficient due to the satellite geometry. A possible candidate for this effect is the fourth-stage rocket nozzle with its tapered sides and open end. Owing to the complicated geometry, it is difficult to estimate the magnitude of the effect; however, it seems reasonable to expect that the difference in aerodynamic force will be no more than

$$\Delta F_D = 0.1 A_0 \rho V^2 = 0.05 \text{ dynes} \tag{5}$$

where 0.1 is approximately the ratio of the momentums of the diffusely reflected particles to that of the incident particles, and A_0 is the nozzle end area taken to be of the order of 200 cm^2. Since the nozzle is approximately 150 cm from the center of mass, a maximum torque of 7 dyne-cm seems reasonable. Furthermore, this maximum torque is only effective for approximately half the tumble cycle, and for only about 0.1 of the orbital revolution. Therefore, the net torque appears to be less than 0.35 dyne-cm.

3. *Torque Due to Permanent Magnetic Moment*

A magnetic moment of 1 amp-meter^2 in the geomagnetic field taken to be approximately 2×10^{-5} weber/meter^2 experiences a torque of 200 dyne-cm. These figures are representative of the effects observed on Tiros I by Bandeen and Manger of Radio Corporation of America (RCA) [9]. However, for a tumbling satellite like Explorer XI, the permanent dipole moment along the longitudinal axis tends to be averaged to zero by rotational symmetry in much the same manner as the aerodynamic torques.

To examine this effect more closely, a careful integration technique was developed by Cunningham of Berkeley [10], and the Marshall Space Flight Center Computation Division under the supervision of C.A. Lundquist [11]. The Lagrangian equations of motion were written using the Euler quaternion parameters [12] as the generalized coordinates to express the satellite orientation. Since there are more coordinates than degrees of freedom, it is necessary to introduce a Lagrange undetermined multiplier to supply the additional constraint equation. The reason for choice of the Euler quaternion parameters in lieu of the more conventional Euler angles was that the parameters are symmetric and real (unlike Cayley-Klein parameters which are complex); they range in value from -1 to $+1$; they do not involve trigonometric functions; and, unlike the Euler angles, they are well-defined for all orientations.

For these reasons, the use of the Euler parameters is ideal for a fast and accurate fixed-point integration routine.

From this program it was found that the first-order torques due to the longitudinal component of permanent magnetic moment did indeed cancel out. There are some second-order torques that arise from the fact that the magnetic field has changed owing to the geographic change of the satellite during one tumble cycle. It was found for the case of Explorer IV that the second-order torques could be on the order of 100 dyne-cm, provided the longitudinal permanent dipole moment was 700 amp-meter^2, which is about 80% saturation. Such a dipole moment would produce a maximum instantaneous torque of 3.50×10^5 dyne-cm in a magnetic field of 5×10^{-5} weber/meter^2. However, it is seen that the net torque is on the order of 10^{-3} of the maximum instantaneous torque. No measurement was made on Explorer IV to determine the magnetic moment prior to launch. A crude measurement of the permanent magnetic moment for Explorer XI was made, however, and it was found that the longitudinal component was approximately 0.6 amp-meter^2. This can produce a maximum instantaneous torque of 300 dyne-cm, and from the Explorer IV computation results it appears that the net torque would be negligible.

In the Explorer XI magnetic moment measurement, it was found that there also existed a tranverse component of magnetic moment estimated to be 0.03 amp-meter^2. Since Explorer XI rotated about an axis of maximum moment of inertia with no longitudinal roll, a component of this magnetic moment could have a constant projection on the tumble axis and produce a first-order torque. The maximum torque from this appears to be only 15 dyne-cm.

4. *Torque Due to Eddy Currents*

A conductive surface rotating in a magnetic field experiences induced eddy currents that produce a torque which tends to oppose the rotation and remove angular momentum from the body. Owing to the current path geometry, such a torque may have a component perpendicular to the angular momentum, and hence cause a precession [13]. The calculation of the exact precession component is complicated by the irregular geometry, but certainly one would expect the order of magnitude of such a torque to be no more than the damping torque. The damping torque can be determined by observing the loss of angular momentum resulting in an increase in tumble period from 12.42 sec to 14.62 sec in 96 days. This corresponds to a damping torque of 1.8 dyne-cm. Again it is seen that the precessing torque is considerably smaller than is required to explain the observed precession.

5. Torque Due to Induced Magnetic Moment

It was suggested by Colombo [14] of the Smithsonian Institute that a magnetic moment induced in the ferromagnetic portions of the satellite by the geomagnetic field might be responsible for the observed torques in Explorer IV. This possibility has the desirable feature that the induced moment is essentially along the axis of symmetry, but reverses polarity during a tumble cycle. This allows these torques to add rather than cancel as in the case of the permanent dipole moment.

Explorer XI is constructed primarily from aluminum, which is diamagnetic. However, the attached final stage casing is 410 stainless steel. An order-of-magnitude estimate of the induced magnetic moment could be made if the permeability of the stainless steel was known for very small values of flux density in the ferrous material. Such information is not readily available in the standard handbooks. Also, the effect due to any premagnetization must be considered. The available permeabilities on 410 stainless have values of 110–180 at $B=200$ gauss and 600–1000 at $B = 6000$–7000 gauss. Note that the value of 110 is at 200 gauss which corresponds to an applied field of 2 oe. All that can be said is that the permeability is probably much less than 110 for the geomagnetic field value of 0.2 oe. Therefore, it appears that a meaningful order-of-magnitude calculation of this effect is difficult to make.

As an alternative, we at Marshall Space Flight Center attempted to measure the induced magnetic moment by timing the period of oscillation of a small bar magnet near the end of an old rocket casing. This was done with the casing oriented in a north-south and then in a south-north direction. After the north horizontal component of the earth's field (0.22 oe) was subtracted, no detectable difference was noted in the flux density regardless of orientation. This seems to indicate that the field near the test magnet was due entirely to the permanent magnetic moment in the rocket casing and the geomagnetic field. The statistics of our data indicated that an induced moment greater than 0.02 amp-meter^2 should have been detected.

If a maximum induced moment of 0.02 amp-meter^2 results from an applied field of 0.22 oe or flux density of 0.22×10^{-4} webers/meter^2,

$$\mathbf{M} = \frac{\mu - 1}{\mu_0} V(\mathbf{B} \cdot \mathbf{A})\, \mathbf{A} = 0.02 \text{ amp-meter}^2 \tag{6}$$

$$\frac{\mu - 1}{\mu_0} V = 909 \, \frac{\text{amp-meter}^4}{\text{weber}} \tag{7}$$

where $\mathbf{A}$ is a unit vector along the longitudinal axis.

The maximum resultant torque from a flux density at 5×10^{-5} webers/meter^2 is

$$\mathbf{L} = \mathbf{M} \times \mathbf{B} = 11 \text{ dyne-cm} \tag{8}$$

Averaging over a tumble cycle reduces the torque to a maximum of 6 dyne-cm.

6. Other Possible Sources of Torques

An often-cited source of external forces on satellites, particularly in the case of Echo, is radiation pressure. Since the radiation power density is 1358 watts/meter^2, the resulting force is 4.52×10^{-5} dynes/cm^2. This produces a maximum force of 0.2 dynes on Explorer XI. Since the displacement of the center of area from the center of mass is approximately 15 cm, this results in a torque of 3 dyne-cm. The same objection to aerodynamic torques also applies in this case; the net torque is canceled by rotational symmetry.

Meteorite impacts can be ruled out on the grounds that they are randomly occurring events, whereas the observed torques appear to be continuous. Furthermore, the momentum transferred by meteoric collisions appears many orders of magnitude too small to account for the observed effects.

Differences in the drag cross section due to electrostatic effects have been considered. Since the satellite is a conducting cylinder moving through a magnetic field, an electric field of approximately 0.4 volts per meter is induced across the longitudinal axis of the cylinder. This produces a charge separation that may influence the impact parameters of incident ions. Thus, the negative end will appear to have a larger drag cross section than the positive end to the positively charged atmospheric ions. The interesting feature of this mechanism is that each end alternates sign during a tumble cycle so that the effect does not cancel owing to rotational symmetry.

The Hamiltonian for an incident particle may be written:

$$H = \frac{p_r^2}{2m} + \frac{p_\theta^2}{2mr^2} + \frac{qq'}{4\pi\epsilon_0 r} \tag{9}$$

where q' is the charge of the incident particle. When r is very large,

$$H = \frac{p_r^2}{2m} = \frac{mv^2}{2} \tag{10}$$

Let r' be the impact parameter; then

$$p_\theta = mvr' \tag{11}$$

The closest approach the particle makes to the cylinder is found by requiring $p_r = 0$. This results in

$$\frac{r'^2 - r^2}{r} = \frac{-2qq'}{4\pi\epsilon_0 mv^2} \tag{11}$$

Let the closest approach be the radius of the cylinder, r_0. Let the impact parameter r' for particles that will just graze the cylinder be

$$r' = r_0 + \Delta r \tag{12}$$

and assume $\Delta r \ll r_0$. The change in impact parameter due to charge q in the cylinder is

$$\Delta r = \frac{-qq'}{4\pi\epsilon_0 mv^2} \tag{13}$$

The q in a small increment of the cross section of the cylinder at distance x from the origin is

$$V(r_0) = \frac{qq'}{4\pi\epsilon_0 r_0} = \Phi(x)\, q' \tag{14}$$

hence,

$$\frac{q}{4\pi\epsilon_0 r_0} = \Phi_{(x)} \tag{15}$$

Since

$$\nabla\Phi_{(x)} = -\mathbf{E} \tag{16}$$

for the one-dimensional case,

$$\Phi_{(x)} = -Ex \tag{17}$$

At the ends of a cylinder 2 meters long and 0.075 meter in radius (approximating Explorer XI), the Δr may be calculated assuming

$$q' = 1.6 \times 10^{-19} \text{ coulomb}$$

$$x = l/2 = 1 \text{ meter}$$

$$r_0 = 0.075 \text{ meter}$$

$$E = |\mathbf{V} \times \mathbf{B}| = 0.4 \text{ volts/meter} = 0.4 \text{ joule/coulomb-meter}$$

$$m = 22 \times 1.6 \times 10^{-27} \text{ kg}$$
(assuming 22 to be the average molecular weight)

$$v^2 = 6.4 \times 10^7 \text{ meter}^2/\text{sec}^2$$

The resulting Δr results in making the positive end approximately

23 cm^2 smaller in cross section and the negative end 23 cm^2 larger for the ionized atmospheric particles. If we assume the maximum density of ionized particles to be 2×10^6 ions/cm^3 [15], the net torque becomes only 0.06 dyne cm.

IV. Conclusions

After various mechanisms have been considered in an attempt to interpret the observed torques in Explorer XI, a completely satisfactory explanation is yet to be found. It is quite possible that erroneous or oversimplified assumptions have been made in some cases, and the effect may have been lost in such assumptions. The fact remains that the satellite angular momentum does indeed change its orientation and that some mechanism must supply external torques to account for this effect. It seems that all the obvious mechanisms have been considered in this study, so either invalid assumptions have been made, or some less obvious effect is responsible. To check the former possibility, a study is under way to compute the directions of the torques supplied by the more promising mechanisms, and compare them with the directions of the observed torques without regard to magnitude. Perhaps such a study will indicate which effect warrants more careful analysis.

One possible effect that needs more careful consideration is the torque produced by a permanent transverse magnetic moment. The transverse magnetic moment has been estimated at 0.03 amp-meter^2 from a very crude measurement. If this estimate is small by an order of magnitude, a sufficient torque will result to explain the observed precession. It appears difficult to measure accurately the transverse magnetic moment, but an analysis of the directions of such a torque may indicate whether this is the responsible mechanism. If, indeed, the transverse permanent magnetic moment seems to explain Explorer XI observations, there may still be some difficulty in explaining Explorer IV results by the same mechanism. This difficulty arises because Explorer IV is known to have had a residual roll about its longitudinal axis, whereas Explorer XI has no such roll. Therefore, in Explorer IV such torques would tend to average to zero.

References

1. G. Heller, "Problems Concerning the Thermal Design of Explorer Satellites," *IRE Trans. on Military Electron.* **4**, 109 (1960).
2. J. A. Van Allen, C. E. McIllwain, and G. H. Ludwig, Radiation observations with satellite 1958 Epsilon. *J. Geophys. Res.* **64**, 279 (1959).
3. R. J. Naumann, Recent information gained from satellite orientation measurement. *Planetary and Space Sci.* **7**, 445-453 (1961).

4. C. A. Lundquist, R. J. Naumann, and S. A. Fields, Recovery of further data from 1958 Epsilon. "Space Research," (Jager, Moore, and Vande Hulst, eds.), Vol. II. North-Holland Publ. Co., Amsterdam, 1962.
5. C. A. Lundquist, R. J. Naumann, and A. H. Weber, Directional flux densities and mirror point distribution of geomagnetically trapped charged particles from satellite 1958 Epsilon measurement. *Am. Geophys. Union 1st Western Natl. Meet. UCLA, December 27-29, 1961*; submitted for publication in *J. Geophys. Res.*
6. R. J. Naumann, S. A. Fields, and R. L. Holland, Explorer XI (S-15) Orientation Analysis, *NASA Tech. Note D-1117.* Marshall Space Flight Center, Huntsville, Alabama, 1962.
7. R. E. Roberson, and D. Tatistcheff, The potential energy of a small rigid body in the gravitational field of an oblate spheroid. *J. Franklin Inst.* **262**, 209-214 (1956).
8. B. F. Doolin, Gravity Torques on an Orbiting Vehicle, *NASA Tech. Note D-70.* Ames Research Center, Moffet Field, California, 1960.
9. W. R. Bandeen and W. P. Manger, Angular motion of the spin axis of Tiros I meteorological satellite due to magnetic and gravitational torques. *J. Geophys. Res.* **65**, 2992-5 (1960).
10. L. E. Cunningham, Research on the Motions of Artificial Satellites. Contract No. NAS 8-2514, University of California, 1962.
11. C. A. Lundquist and R. J. Naumann, Orbital and rotational motion of a rigid satellite. *Seminar Proc. Tracking and Orbit Determination.* Jet Propulsion Lab., Calif. Inst. Technol., Pasadena, Calif., 1960.
12. E. T. Whitaker, "A Treatise on the Analytical Dynamics of Particles and Rigid Bodies," 4th Ed., Dover, New York, 1944.
13. W. R. Smythe, Damping and Precessing Torques on Spinning Prolate Spheroid with Skin Effect. OOR Contract No. DA-36-034-ORD-1535 with Duke University, 1957.
14. G. Colombo, The Motion of Satellite 1958 Epsilon around Its Center of Mass, *Research in Space Sci. Smithsonian Astrophys. Obs. Spec. Rept. No. 70,* July 18, 1961.
15. J. S. Nisbet, Electron Densities in the Upper Ionosphere from Rocket Measurements, *Penn. State Univ. Ionospheric Res. Sci. Rept. 126,* December 10, 1959.

Note: Since this paper was prepared, this problem has been treated in more detail; the observed results are explained by considering a permanent magnetic dipole moment in the satellite shell. See NASA Rept. TTR-183 (Marshall Space Flight Center) 1963.

Horizon Sensing in the Infrared: Theoretical Considerations of Spectral Radiance

D. Q. WARK and J. ALISHOUSE

U. S. Weather Bureau, Washington, D.C.

and

G. YAMAMOTO

Tohoku University, Sendai, Japan

I. INTRODUCTION

THE INFRARED RADIANCE of the earth and atmosphere toward space varies spectrally and with zenith angle according to the temperature distribution and the absorption by the atmospheric gases. As viewed from an artificial satellite, a radiometer of narrow angle of view can measure the radiance integrated over a spectral interval appropriate to the requirement.

Measurement of radiance at the limb of the earth has been used to establish the horizon for the purpose of stabilizing satellites. A horizon sensor should be sensitive in a spectral region having as little contrast over the disk as possible, and it must not be subject to uncertainties in the horizon itself which are greater than the degree of stabilization required. For example, sensors sensitive to reflected solar radiation would be unsuitable because of the contrast between clouds and most surfaces, and sensors sensitive to some infrared spectral intervals would view areas, such as cloud tops and ground, in which the contrast in radiance was large by virtue of the large temperature differences.

As a means of providing an insight into the variation of radiance across the disk of the earth, calculated values can be determined from atmospheric models which represent the range of probable conditions over different latitudes, seasons, and meteorological conditions. This variation (limb darkening or limb brightening) varies with the spectral region and the atmospheric model, but some regions of the spectrum should exhibit greater uniformity than others, and would therefore be preferable for horizon sensors.

In a previous study [1], the spectral radiance was calculated for many atmospheric models. Because the purpose of the study was to determine limb darkening only for the purpose of transforming radiance to radiative flux, it was not necessary to extend the calculations out to the limb, where the contribution to flux is very small. Instead, a plane parallel atmosphere was used, with zenith angles limited to 78.5 deg; beyond that angle an extrapolation was sufficient. In considering horizon sensors, however, the exact value of the radiance is needed at the limb. At extreme angles the plane parallel atmosphere becomes a poor approximation, and the curvature of the earth and atmosphere must be taken in account. Refraction also becomes significant at large angles, and, although it has a minor effect upon the results, should be included in the calculations.

Several studies have been made of the spectral radiance near the limb [2–6]. The work reported here supplements the other studies by employing narrower spectral intervals, which may be combined with sensor response functions to give relative sensor output, by consideration of atmospheric refraction, or by the use of more recent gas transmittance values.

II. Path Length in the Curved, Refractive Atmosphere

In this investigation both the curvature of the atmosphere and the refraction by air are taken into account. The earth is assumed to be a sphere of radius 6371.0 km, and 0.1 mb pressure is assumed to be the top of the atmosphere. For each atmospheric model used in the calculations, the pressure, humidity, and ozone concentration are given as functions of vertical height Z calculated from the hydrostatic and ideal gas relations; the atmosphere is assumed to be homogeneous in the lateral directions. The refractive index of air n is basically a function of air density, and is given by

$$n = 1 + \frac{CP}{T} \qquad C = 77.526 \times 10^{-6} \frac{^\circ\mathrm{K}}{\mathrm{mb}} \tag{1}$$

where P and T are the pressure and temperature of the air respectively. Hence the refractive index is also given as a function of height in this investigation and will be denoted by n_z. The spectral dependence of n is small enough in the infrared to be ignored.

In calculating the geometric path it will be convenient to consider the following two cases separately: (1) the path which intersects the earth's surface and (2) that which passes entirely above the surface. The geometry is shown in Fig. 1.

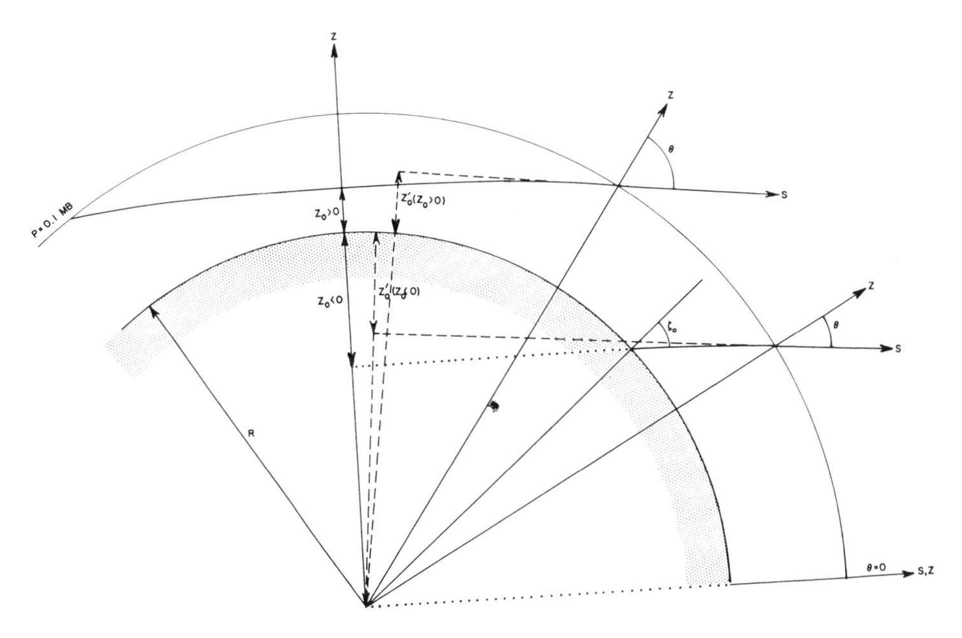

FIG. 1. Geometry of the optical path through the atmosphere. The direction of the ray is indicated by s; Z_0 is the minimum height of the refracted ray, or the minimum projection of its intersection with the earth onto a radius; θ is the zenith angle of the emerging ray at 0.1 mb; Z_0' has the same meaning as Z_0, but for an unrefracted ray; ζ (not shown) is the zenith angle of the ray at any height Z, varying from the surface value ζ_0 to the 0.1-mb value θ.

A. The Path Which Intersects the Earth's Surface ($Z_0 \leqslant 0$).

The zenith angle of the path is ζ_0 at the surface, ζ at any vertical height Z above the surface, and θ at the 0.1-mb level. The vertical height of the 0.1-mb level is given by H, and the radius of the earth by R; n_0, n_z, and n_H are the indices of refraction at the heights indicated by the subscripts. These quantities are related by the well-known relationship

$$\begin{aligned} Rn_0 \sin \zeta_0 &= n_z(R + Z) \sin \zeta \\ &= n_H(R + H) \sin \theta \end{aligned} \tag{2}$$

The element of length along the path is given by

$$ds = \frac{dZ}{\cos \zeta} \tag{3}$$

which becomes, by referring to Eq. (2) and utilizing

$$\sin \zeta_0 = \frac{R + Z_0}{R} \qquad \text{for} \qquad Z_0 \leqslant 0 \tag{4}$$

$$ds = \frac{(R + Z)\, n_z\, dZ}{[(R + Z)^2\, n_z^{\,2} - n_0^{\,2}(R + Z_0)^2]^{1/2}} \tag{5}$$

B. The Path Which Passes through the Atmosphere ($Z_0 > 0$)

To calculate the geometric path in this case it is necessary simply to change n_0 to n_{z_0}, the index of refraction at the height z_0, in Eq. (5). The resulting formulas corresponding to Eqs. (2) and (5) are then:

$$(R + Z_0)n_{z_0} = (R + Z)\, n_z \sin \zeta = (R + H)\, n_H \sin \theta \tag{6}$$

$$ds = \frac{(R + Z)\, n_z\, dZ}{[(R + Z)^2\, n_z^{\,2} - (R + Z_0)^2\, n_{z0}^2]^{1/2}} \tag{7}$$

The limb darkening will be shown by comparing the radiances of parallel beams leaving the atmosphere. In doing so it will be convenient to take the zenith angle at the top of the atmosphere, θ, as a parameter in specifying the rays. However, the paths are calculated by taking ζ_0 or Z_0 as a parameter, and rearranging them in order of θ. The critical case between A and B is the path which touches the earth surface tangentially; in the case of the standard atmosphere, for example, the critical value of θ is 81 deg 46.7 min.

III. Transmittance

The division of the spectrum into 77 intervals covering the infrared regions from 0 to 2330 cm^{-1} and the division of the atmosphere into 200 layers in calculating the outgoing intensity of radiation were made as in the previous work [1]; the transmittance values used in this work are also determined in essentially the same way as in the previous work. In the present case, however, the optical path length is much longer—the longest path is about 8 times that in the previous case—so that the transmittance values must be extrapolated for the longer path, on which no measurements have been made. There was not much difficulty in making the extrapolation, except in several intervals of the water vapor bands which lie near the window region. It was found that the mere extrapolation of the transmittance curves in and adjacent to the window region leads to the incongruity of greater absorption in the window than in the adjacent regions of the water vapor bands at the greater path lengths. The transmittance curves for the window region used in the previous work are based on reference [7] and are approximately exponential curves, while the transmittance curves for the rest of the water vapor band used in the previous work are based on Cowling's [10] universal transmittance, which agrees with Goody's [8] random model transmittance.

It is evident that the puzzling result described above can be ascribed to using such different transmittance curves for neighboring spectral regions. One may conjecture that the character of the transmittance of the region adjacent to the so-called window region is in a transition between that of the window region and that of the strongly absorbing region, which is expressed by Cowling's [10] curve. In these intermediate regions the line intensities become much greater than in the window region, although at the same time a continuous background absorption, due to the wings of lines in the strongly absorbing region, is more pronounced in the intermediate region than in the window region. Hence the transmittance of the intermediate region is a superposition of the line absorption and continuous background absorption, with more emphasis on the line absorption than in the window region.

Based on the above reasoning for the intermediate regions from 1232 to 1300 cm^{-1} of the 6.3 μ band and from 600 to 778 cm^{-1} of the rotation band, transmittance curves have been developed. For the window region extrapolations of the approximately exponential curves are used as in the previous work [1]. Cowling's [10] universal transmittance curves are used in this investigation for the strongly absorbing regions from 1300 to 2160 cm^{-1} in the 6.3 μ band and from 0 to 600 cm^{-1} in the pure rotation band.

Except in the window region, optical path length is calculated from the local density ρ integrated over the geometric path s:

$$u = \int_0^s \rho \, ds \tag{8}$$

The effective pressure P_e is given by

$$P_e = \frac{\int_0^u P\,du}{P_0\,u} \tag{9}$$

where P_0 is standard pressure. In addition, effective temperature is used in the carbon dioxide band, calculated from

$$T_e = \frac{\int_0^P T\,dP}{\int_0^P dP} \tag{10}$$

In the window region, the effective path length u_e may be obtained from

$$u_e = \int_0^S \rho \frac{P}{P_0}\,ds \tag{11}$$

Tables of transmission have been prepared in terms of the above quantities.

IV. Calculation of Spectral Radiance

The radiance has been calculated from the integral form of the radiative transfer equation for each of the 77 spectral intervals; about 15 values of Z_0 were used for each model atmosphere to show the variation of radiance across the limb.

Integration was performed numerically by dividing the atmosphere into 200 concentric spherical shells extending from the surface to 0.1 mb, and substituting a summation:

$$I_{\bar{\nu}}(Z_0) = B_{\bar{\nu}}(T_{0.1}) + \sum_{i=1}^{200} \tau_{\bar{\nu}_i}(Z_0) \left[\frac{\partial B_{\bar{\nu}}(T)}{\partial T}\right]_{\bar{T}_i} [\Delta T]_i \tag{12}$$

where $I_{\bar{\nu}}(Z_0)$ is the radiance in a spectral interval centered at frequency $\bar{\nu}$, $T_{0.1}$ is the temperature at 0.1 mb, $\tau_{\bar{\nu}}$ is the mean transmittance over the interval, and $B_{\bar{\nu}}$ is Planck radiance at $\bar{\nu}$. Integrations shown elsewhere in this paper were performed by similar summations over the 200 layers.

Calculations were made for atmospheric models representing a wide range of radiance conditions:

(a) The ARDC (1959) model atmosphere, with clear sky.

(b) Albuquerque, New Mexico, 0000 GCT, July 11, 1958, clear with the surface temperature 311°K and the tropopause 200°K at 100 mb.

(c) Ponape, Caroline Islands, 1200 GCT, May 17, 1958, with an assumed thick overcast top at 100 mb, where the temperature was 187°K.

(d) Resolute, Northwest Territories, Canada, 1200 GCT, Dec. 31, 1958, with an assumed thick overcast top at 400 mb, where the temperature was 221°K.

The carbon dioxide mixture was assumed to be uniform at 0.2479 cm atm/mb (0.031% by volume), and ozone mixture was consistent with the results of Dütsch [9]. Water vapor mixtures in the troposphere were normal, a constant frost point was used in the stratosphere to 10 mb, and above that a constant mixing ratio was employed.[1]

V. Results

The variation of spectral radiance in several narrow intervals, in the "window" and in each of the principal infrared bands, is shown in Fig. 2. The result for each atmospheric model is given in each of the five spectral intervals. The ordinate is radiance, and the abscissa is Z_0', the apparent value of Z_0 from outside the atmosphere (see Fig. 1).

Figure 2(a) shows one of the most transparent parts of the "window," 908–948 cm^{-1} (10.5–11.0 μ). As one might expect, there is a rapid drop of radiance beyond $Z_0' = 3$ km (which corresponds to $Z_0 = 0$). The amount of limb darkening for each atmosphere is not extreme between the center of the disk ($Z_0' = -6371$ km) and $Z_0' = 3$ km. However, a great range of radiances (300 to 5100) is found over the disk for even this limited sample of atmospheres.

Figure 2(b) shows the region of the ozone band, 948–1100 cm^{-1} (9.1–10.5 μ). Here the limb darkening (or limb brightening in one case) between the center of the disk and $Z_0' = 3$ km is somewhat greater. In addition, there is a hump for each atmosphere, centered at about $Z_0' = 25$ km; this is caused by the long optical path through the most dense ozone regions, where the temperature is also somewhat higher than the tropopause minimum. Here, as in the "window," the range of radiances is very large over the disk (1100 to 13,200).

[1] For additional details, see "Calculations of the Earth's Spectral Radiance for Large Zenith Angles," *Meteorological Satellite Laboratory Report No. 21*, available from the authors upon request.

Figure 2(c) shows one of the most absorbing regions of the 6.3 μ band of water vapor, 1500–1525 cm^{-1} (6.56–6.67 μ). The erratic behavior of the radiance is caused by the different distributions of water vapor and temperature, and is accented by the extreme variation of radiance with temperature at these wavelengths (about T^8). Although the ultimate limb at about $Z_0' = 50$ km is similar in all cases, the variations over the disk are rather large (24 to 105).

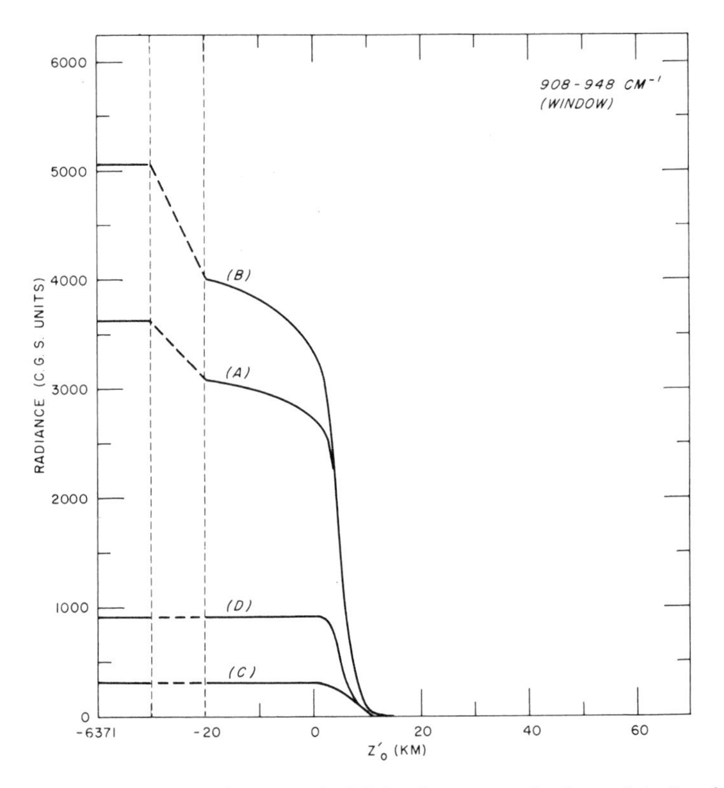

FIG. 2a. Variation of radiance with Z_0' for four atmospheric models in the 908–948 cm^{-1} interval in the "window." The radiance for the vertical beam ($Z_0' = -6371$ km) is shown at the left; the radiance near the limb is to the right of the dashed lines.

Figure 2(d) shows an interval in the rotational band of water vapor at 300–325 cm^{-1} (30.8–33.3 μ). The limb is similar in each model, there is comparatively modest limb darkening or limb brightening, and the range of radiance over the disk is more limited (940 to 1600). Although the absorption coefficients in this interval are about the same as in the 1500–1525 cm^{-1} interval, the radiance is more uniform by

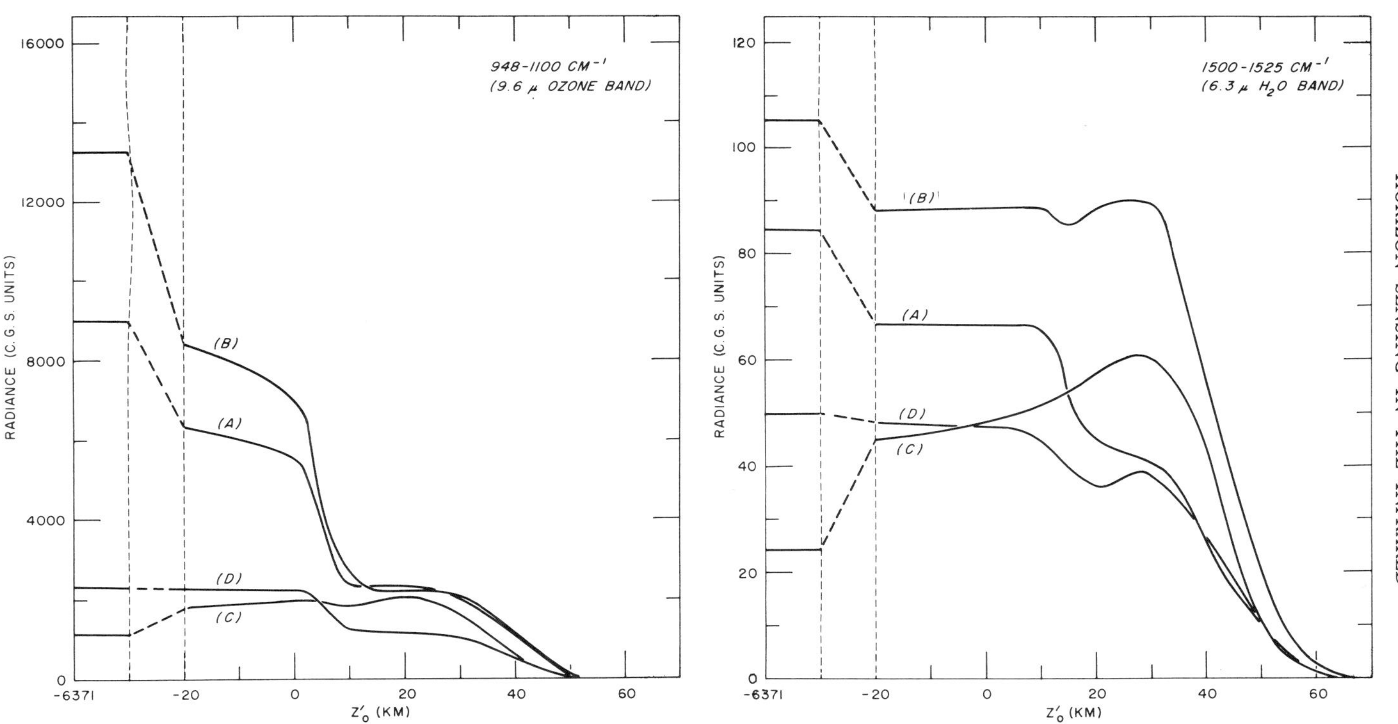

FIG. 2b. Variation of radiance in the 948–1100 cm^{-1} interval in the 9.6 μ ozone band.

FIG. 2c. Variation of radiance in the 1500–1525 cm^{-1} interval in the 6.3 μ water vapor band.

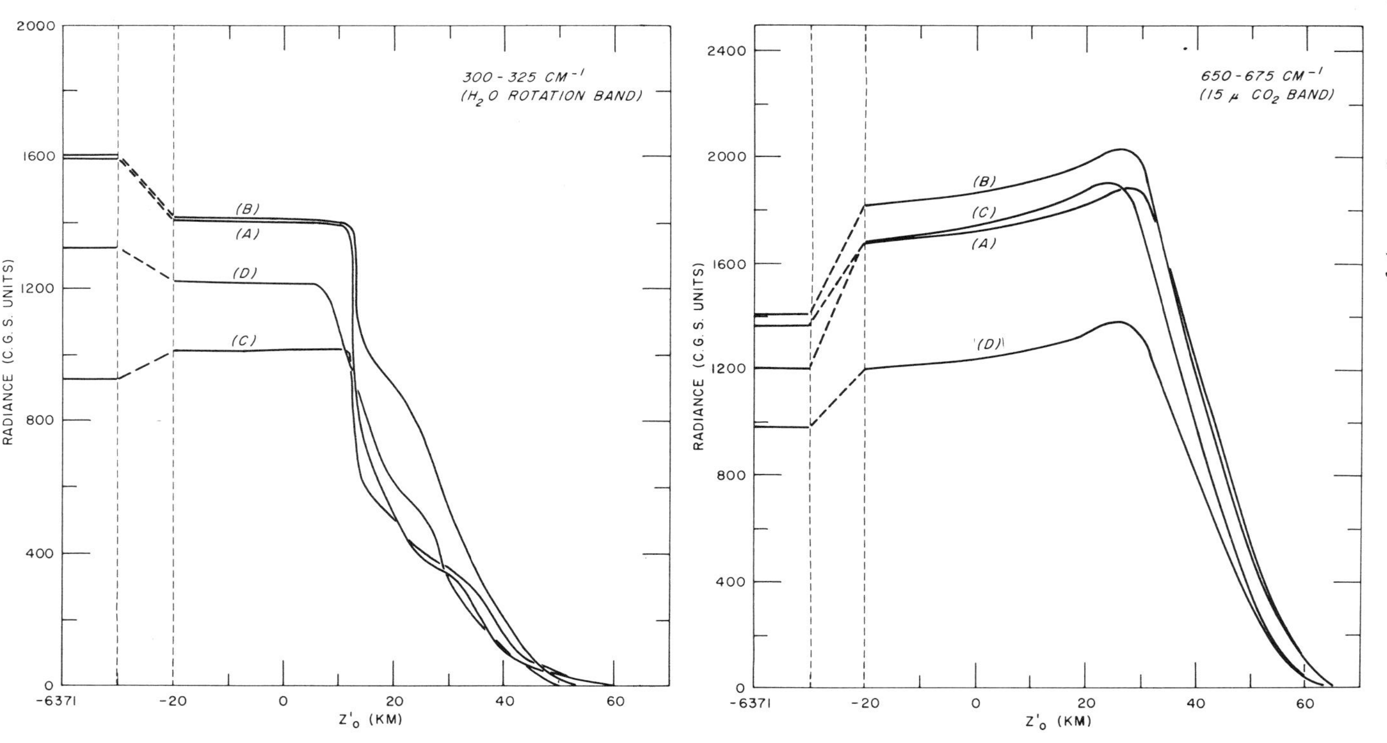

FIG. 2d. Variation of radiance in the 300–325 cm^{-1} interval in the water vapor rotation band.

FIG. 2e. Variation of radiance in the 650–675 cm^{-1} interval in the 15 μ carbon dioxide band.

virtue of the lower sensitivity to temperature (about T^2) in this part of the spectrum.

Figure 2(e) shows the interval 650–675 cm^{-1} (14.8–15.4 μ), which is the most absorbing part of the carbon dioxide band. The range of radiances is not extreme (980–2030), and the ultimate limb at about $Z_0' = 50$ km exhibits the most consistent behavior of any of the regions examined. It will be noted that there is limb brightening for each model atmosphere.

Of the five spectral regions examined, only two, the carbon dioxide and the rotational water vapor band, exhibit characteristics desirable for horizon sensors. These two show the least variation over the disk (range of values of the radiance) and consistency in the ultimate limb. However, these intervals are not necessarily representative of horizon sensors. To make the results of this study more practical, the inner product of a filter function ϕ_ν and the radiance is computed to give the filtered radiance

$$I_\phi(Z_0') = \sum_{i=1}^{n} [I_{\bar{\nu}}(Z_0')\,\phi_{\bar{\nu}}\Delta v]_i \tag{13}$$

as seen by the sensor. The filter functions, normalized to central values of 1.0 but otherwise typical of current technology, are shown for the two bands in Fig. 3.

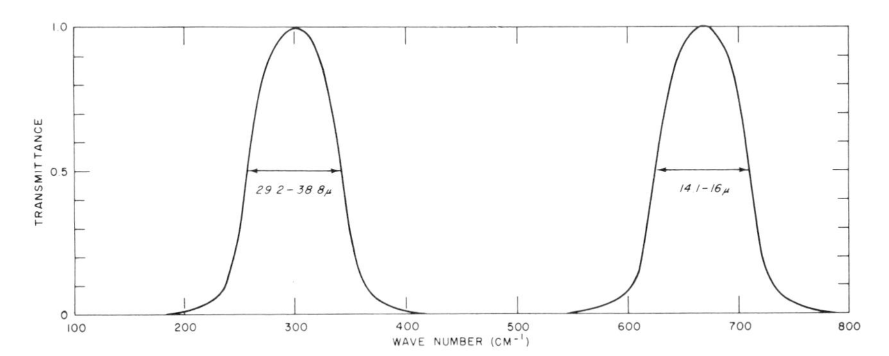

FIG. 3. Filter transmittance in the 15 μ carbon dioxide band and in the rotational water vapor band.

Figure 4(a) shows the filtered radiance in the carbon dioxide band. The range of values over the disk is now less than a factor of 2. The limb brightening is extremely small, except in atmosphere C, which has an extremely high and cold overcast not normally encountered; even so, this case is not seriously disturbing. The ultimate limb is very well behaved.

Figure 4(b), giving the filtered radiance in the rotational water vapor band, is almost indistinguishable from Fig. 2(d). This is because the absorption coefficients over the filter region do not change drastically from those in the 300–325 cm^{-1} interval; this is in contrast with the carbon dioxide band, where the reverse is true.

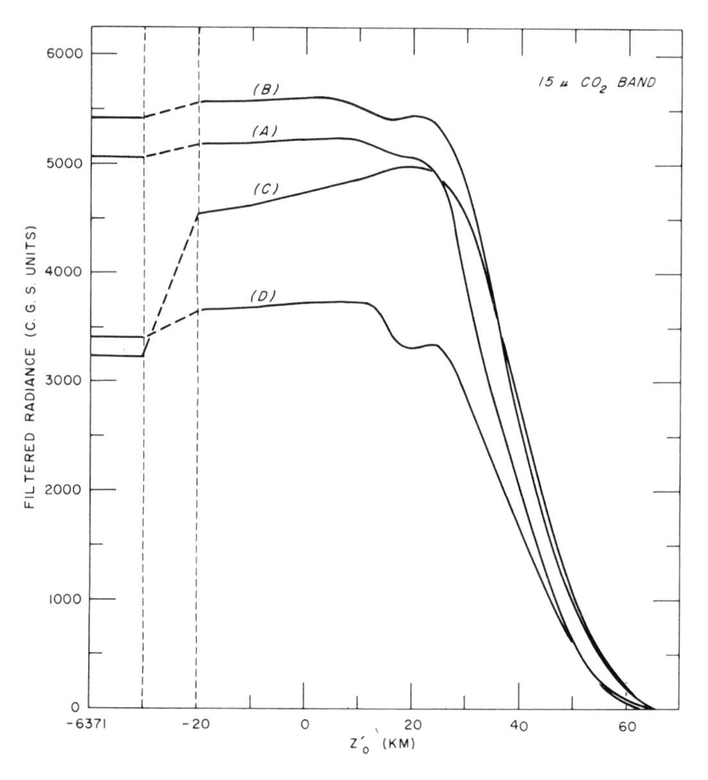

FIG. 4a. Variation of the filtered radiance $I_\phi(Z_0')$ with Z_0' for four atmospheric models in the carbon dioxide band.

Examining Figs. 2 and 4, one concludes that only two spectral intervals are well suited to horizon sensors. These two would appear to be about equally satisfactory: there would be little chance of false horizon indications resulting from the radiance being below the threshold value of the instrument; and the horizon is reasonably similar for all model atmospheres near the ultimate limb.

Instrumental considerations probably dictate the selection of the spectral interval. In the rotation band of water vapor the band pass of the filter could be increased over that shown in Fig. 3 without significant effect on the behavior of the filtered radiance near the limb. In the carbon dioxide band, however, an increase in the band pass would allow a dilution of the well-behaved radiance by the erratic radiance

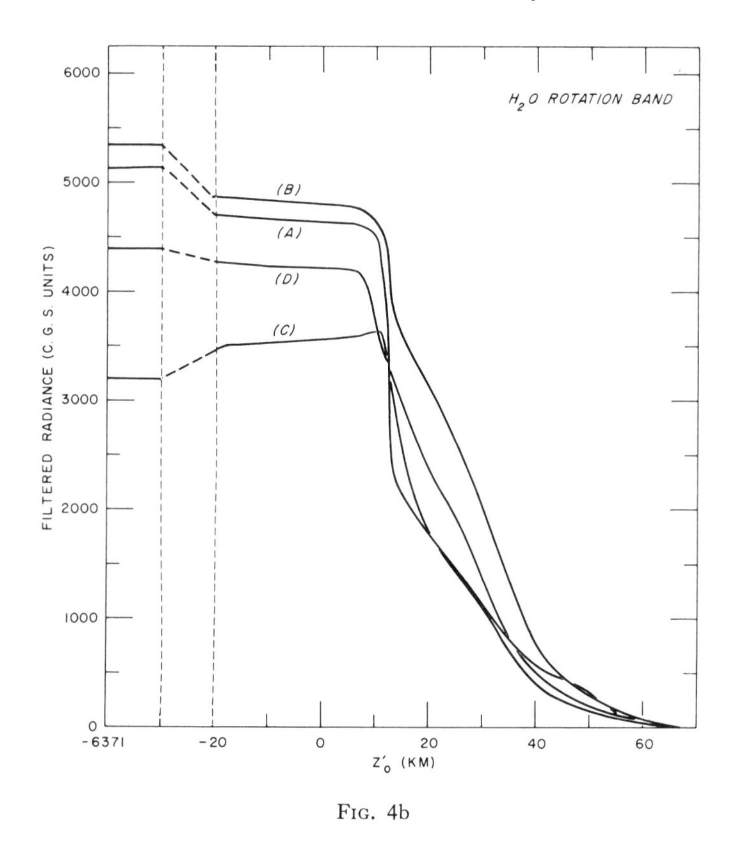

FIG. 4b

found in the window. A broad filter to include both the rotation and carbon dioxide bands would be inferior because of the more transparent region around 500 cm^{-1} (20 μ); however, it might be acceptable, and should not be ignored, in view of the better signal-to-noise ratio achievable at higher levels of radiance.

ACKNOWLEDGMENT

The authors express their thanks to F. VanCleef for writing the computer programs used in this study.

References

1. D. Q. Wark, G. Yamamoto, and J. Lienesch, Methods of estimating infrared flux and surface temperature from meteorological satellites. *J. Atmos. Sci.* **19**, 369-384 (1962).
2. K. Ya. Kondratiev and K. E. Yakushevskaya, The angular distribution of the outgoing thermal radiation in the different regions of the spectrum. *Artificial Earth Satellites* (*Acad. Sci. USSR*) **14**, 13-29 (1962).
3. K. Ya. Kondratiev and K. E. Yakushevskaya, The angular distribution of the outgoing thermal radiation in the different regions of the spectrum. *Proc. Intern. Symp. Rocket and Satellite Meteorol.* (1962) pp. 254-277. (North-Holland Publ. Co., Amsterdam).
4. R. A. Hanel, W. R. Bandeen, and B. J. Conrath, The infrared horizon of the planet earth, *J. Atmos. Sci.* **20**, 73-86 (1963).
5. J. W. Burn, private communication, 1963.
6. E. P. Ertsgaard, private communication, 1963.
7. W. T. Roach and R. M. Goody, Absorption and emission in the atmospheric window from 770-1250 cm^{-1}. *Quart. J. Roy. Meteorol. Soc.* **84**, 319-333 (1958).
8. R. M. Goody, A statistical model for water vapor absorption. *Quart. J. Roy. Meteorol. Soc.* **78**, 165-169 (1952).
9. H. W. Dütsch, Vertical Ozone Distribution over Arosa from Three Years Routine Observations of the Umkehr Effect. *Final Rept. AFGI(f14)-905, Geophys. Res. Directorate, Air Force Cambridge Res. Center.* 24 pp, 1959.
10. T. G. Cowling, Atmospheric absorption of heat radiation by water vapor. *Phil. Mag.* **41**, 109-123 (1950).

Horizon Sensing for Attitude Determination

BARBARA KEGERREIS LUNDE

Goddard Space Flight Center, National Aeronautics and Space Administration, Greenbelt, Maryland

I. INTRODUCTION

IN SPACE NAVIGATION, it is often necessary to determine the angles between the direction to the center of another body and the axes of the vehicle. The direction to the center of the other body may be defined as the direction which is perpendicular to the plane of its horizon. This plane may be determined by measuring the angles between direction of the discontinuity between space and the body of interest and the axes of the vehicle at several points on the horizon.

II. PHYSICAL PHENOMENA WHICH MAY BE USED TO DEFINE THE HORIZON

There are several physical phenomena which make possible a definition of the discontinuity between space and the body of interest—in other words, the horizon.

A. Infrared Radiation from the Earth

If earth is the body of interest, one physical phenomenon which differentiates it from space is the infrared radiation it emits. The earth's horizon may be defined as the sharp gradient of infrared radiation which exists at the limb, or border, between it and outer space.

Since the earth has a fairly uniform temperature this gradient may be used for space navigation under a wide variety of circumstances, whether or not the limb is illuminated by the sun. In other words, it may be used under both day and night conditions. Figure 1 shows the

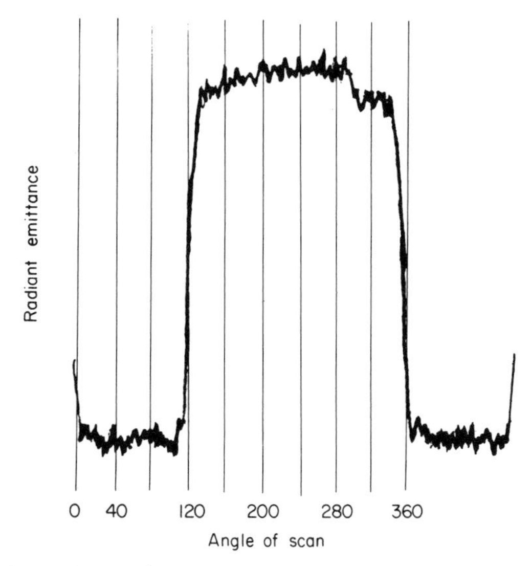

Fig. 1. Scan from Tiros radiometer. Spectral range, 8-12 microns.

output from an infrared sensor which was mounted in Tiros III, a weather satellite. As the satellite rotated, the field of view of the sensor scanned first over the sky, then over the earth, then over the sky again. As its field of view passed over the horizon, the sensor detected the sharp change in the level of infrared radiation which is represented by the sharp rise in the signal. As it passed over the opposite horizon of the earth, it detected the sharp decrease in the level of infrared radiation. It should be emphasized that this figure represents fairly ideal conditions. This sharp gradient of infrared radiation, which exists at the boundary between earth and space beyond, is the physical phenomenon most commonly used at present for sensing the horizon in space navigation.

B. Albedo from the Earth

The sunlight which is reflected by a body in space such as the earth, sometimes referred to as the albedo, is another phenomenon which has been used for horizon determination. However, it is limited in application. At night, when the earth is eclipsing the sun with respect to a space vehicle, no reflected sunlight appears. At other times, the earth appears to have phases like those of the moon, complicating the calculation of the horizon plane by means of the albedo which, incidentally, is defined as the ratio between the light received and the light reflected by a body in space. Figure 2, a colored photograph taken during one of the Mercury suborbital flights, illustrates this phenomenon. The horizon

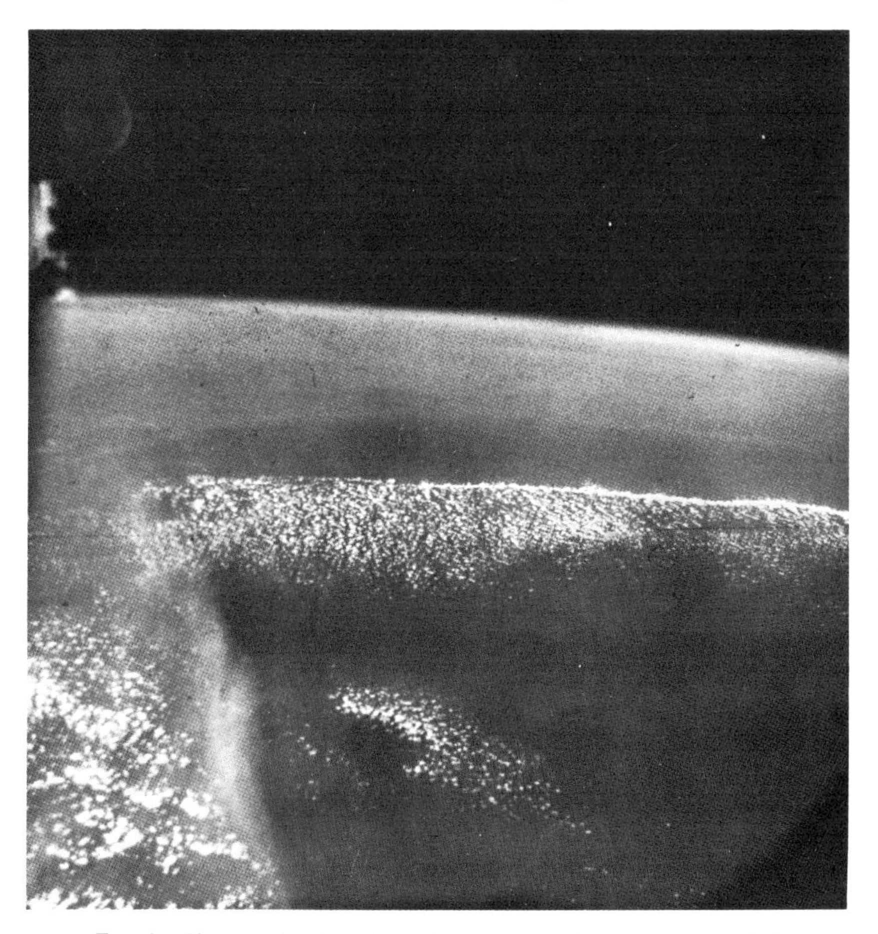

Fig. 2. Photograph taken from Mercury capsule on suborbital flight.

is easily seen as the gradient between the apparent surface of the earth which reflects sunlight, and black space beyond, which does not. The Georgia-Florida coast line is also visible in the picture.

C. Air Glow around the Earth

The upper portion of the earth's atmosphere radiates owing to excitation by the sun. This phenomenon, known as air glow, could be used to define the horizon. The radiation occurs both day and night.

The total visible intensity of this radiation of air glow is approximately equal to the total amount of starlight seen on a clear but moonless night. Air glow is rarely observed on the surface of the earth because it is evenly distributed throughout the sky. Much of the illumination by air glow is in a few spectral lines. Table I shows some experimental data on the

TABLE I

Air Glow Intensity and Altitude of the Sodium 5893 Å Line

Intensity (relative magnitude)		
	Day	Night
Summer	2000	230
Winter	150000	200
Auroral	—	1000
Altitude (two measurements at White Sands, New Mexico)		
87 km		
93 km		

air glow in the sodium line at 5893 Å. This is one of the more intense and better known lines. Its intensity during the day is about seventy-five times the night-time intensity. This air glow radiation comes from a rather narrow band in the atmosphere occurring approximately 90 kilometers above the surface in a layer of about 10 kilometers in thickness. Some scientists think that if the earth were viewed from a long distance away, with a sensor which is sensitive to only one particular band of air glow, a doughnut of radiation would be seen circling a dark earth while the rest of space would be dark, except for point sources. The earth's horizon then could be determined by measuring the physical position of the air glow band. Very few measurements have been made of the air glow phenomenon.

Some of the most recent data indicate air glow's intensity varies from point to point over the earth's surface. Its altitude also appears to vary with time. Altitude measurements have been taken only at White Sands, New Mexico. In Fig. 2, a picture taken during the suborbital Mercury flight, no evidence of the air glow phenomenon is visible. The intensity of air glow in the ultraviolet is only approximately one ten-billionth of the thermal energy the earth radiates. The lack of information on air glow and its low intensity are the main limitations on the usefulness of air glow for space navigational purposes.

D. Other Bodies

The horizons of bodies other than the earth may be defined by use of the same general phenomena already discussed, namely,

1. Emitted thermal radiation;
2. Reflected radiation (albedo); or
3. Other phenomena such as air glow.

The horizons of the nearby planets, Venus and Mars, and perhaps artificial satellites, may be determined by sensing the gradient between the infrared radiation which they emit and that of space.

The visible radiation emitted by the photosphere of the sun may be used to determine its horizon. The horizons of the moon, the near planets, and perhaps artificial satellites also may be determined by the gradient of reflected sunlight at certain times.

III. Scanning Schemes

Any instrument for attitude determination needs, in addition to an intensity sensor, a means for determining the direction of the incident radiation.

This means will be called a scanning scheme.

Five scanning schemes for horizon scanners will be described. Others have been suggested.

A. Passive Scan

The passive scan can only be used on a space vehicle which rotates. A sensor with a small field of view is mounted in a vehicle with its field of view at an angle to the spin axis of the vehicle. As the vehicle rotates, the field of view of the sensor is swept through space.

The data in Fig. 1 were taken with a scanning scheme similar to this. The times at which the field of view of the scanner passes the two horizons can be determined with the use of this data. From knowledge of these time parameters the attitude of the vehicle at any instant can be determined. This technique was employed using infrared sensors on the Tiros weather satellite and on the Goddard Space Flight Center atmospheric structure Explorer XVII satellite. It has also been employed on daytime rocket shots using sensors which detect the sun's reflected light.

B. Conical Scan

A second type of scanning method is the conical scan. Figure 3 is an artist's conception of the Mercury space capsule. The cones represent the fields of view which its horizon scanner sensors sweep out in space. The Mercury capsule has two scanners which are used to determine its pitch and roll errors. Each scanner, again, has a small field of view. This time the rotation of the field of view is done within the scanner.

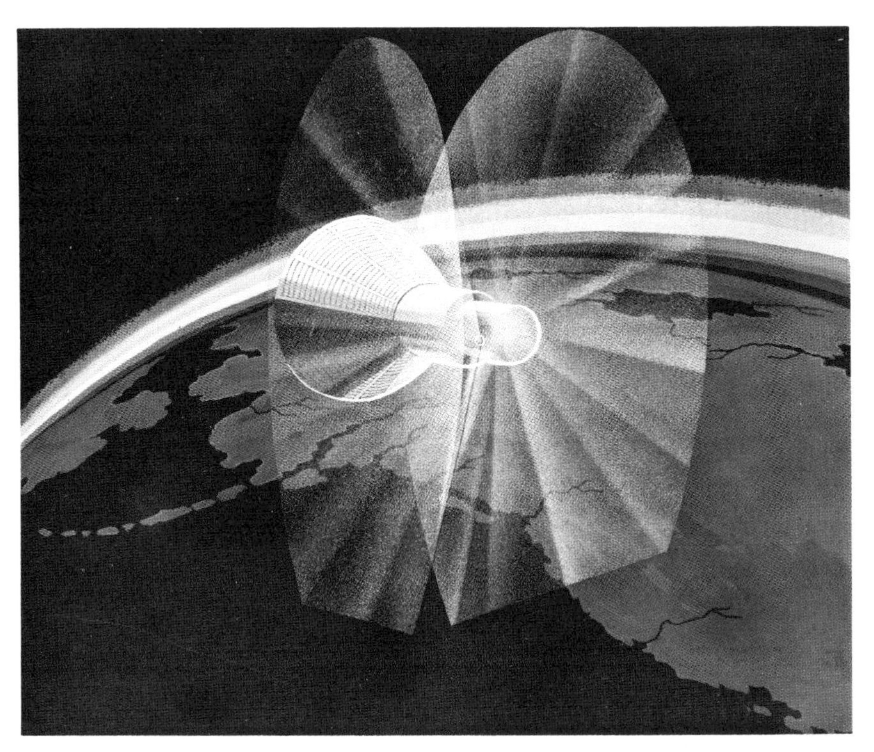

Fig. 3. Conical scanning pattern of horizon scanners on Mercury capsule.

The field of view is scanned in a circle. The cone in space which it scans intersects the earth. The output of the sensor is used to determine the location of the gradient of infrared radiation between space and the earth. The local vertical can then be computed. It is presently planned that the type of scan shown here for the Mercury vehicle will be used in the Nimbus advanced weather satellite. The scan pattern is not very efficient from an information theory point of view. The information as to the direction of the horizon is only a very small part of the information which enters the sensor.

C. Linear Scan

Figure 5 shows a third type of scanning pattern, the linear scan. This is an artist's conception of the Orbiting Geophysical Observatory satellite and the pattern the fields of view of its horizon scanner sensors

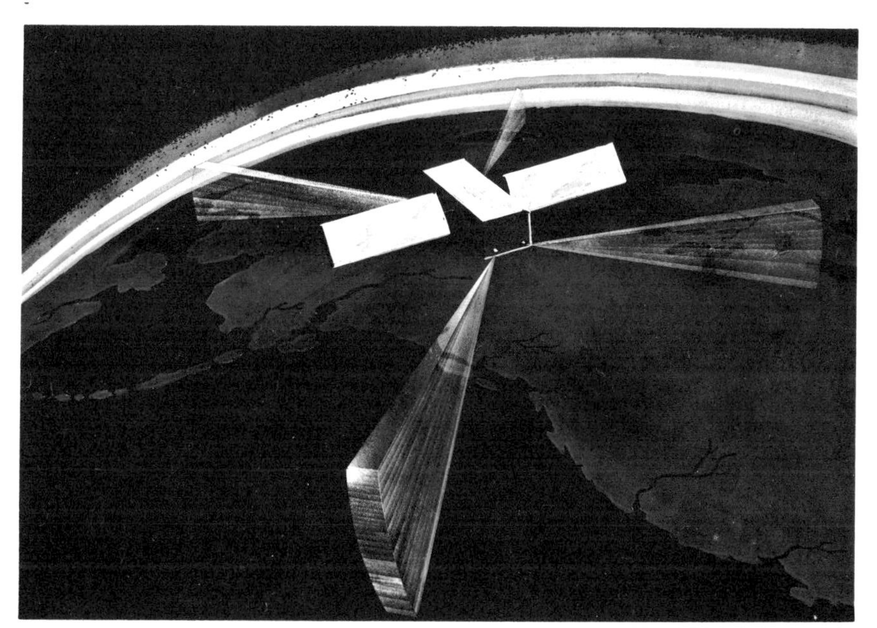

Fig. 4. Linear scanning pattern of horizon scanners on the Orbiting Geophysical Observatory.

sweep out in space. This satellite employs the linear type of scan. In this case, there are four sensors on the vehicle and they each have small fields of view. These fields of view are rotated in a plane from space until they intersect the edge of the earth. They are oscillated at the edge of the earth for the remainder of the time. Once the sensors have locked on the edge of the earth, the angle between the directions of their fields of view and the satellite can be determined. From this information the direction of the local vertical can be calculated. Linear scanning is superior to circular scanning from an information theory point of view because the sensor field of view spends more of its time right at the edge of the earth. Only infrared sensors have been used with these two methods of scanning.

D. Nutating Scan

A fourth type of scanning is called nutating scan. It is similar to a linear scan but there is only one sensor per vehicle. The field of view of the sensor is again small and is initially viewing space. It is swept down until it intersects the horizon. It then oscillates about the horizon. Meanwhile the whole scanning head is rotated about another axis. Thus the projection of the sensor field of view on the earth is a sinusoidal pattern along the earth's horizon. The direction of the local vertical is determined by calculations using the position of the field of view of the sensor with respect to the vehicle at each instant.

E. Illumination Comparison

A fifth type of horizon sensing does not use any moving parts. The illumination from one side of the earth on the scanner is compared with that from the other side of the earth. This information is used to determine the local vertical. For a vehicle within 1000 miles of the earth, it is very difficult to image the whole earth on the scanner because such an imaging system would have to have a very wide field of view. A scanner which uses this method has been proposed. It uses an infrared sensor and a unique imaging system. Another scanner which employs this method was used on the Ranger moon vehicle. It determined the direction to the earth from distances greater than 80,000 miles. It used sensors sensitive to visible light.

IV. Programs Which Use Horizon Scanning

Programs which have used horizon sensors for attitude determination are spinning satellites and rockets including the Goddard Explorer XVII atmospheric structure satellite and the Tiros satellite. The Mercury man-in-space program has used horizon scanners. Horizon scanners have been used to monitor the performance of some rockets. They are also used in the Air Force Agena rocket and satellite system. The Jet Propulsion Laboratory has used an earth sensor in its Ranger spacecraft.

Future programs which have planned to use earth horizon scanners are the Orbiting Geophysical Observatory, the Nimbus advanced weather satellite, the Gemini manned spacecraft, the Apollo manned spacecraft classified programs for the Air Force, and the Saturn rocket.

Future programs for which the earth horizon scanners are under consideration are communication satellites and short-lived rocket probes. Another future use for horizon scanners is on the Mariner spacecraft for trips to Venus and Mars. Horizon scanners would determine the direction between local vertical of those planets and the vehicle axes. Another use for horizon scanners is for the Advanced Orbiting Solar Observatory. This satellite is planned to orbit the earth but observe the sun. An earth horizon scanner could be used to determine the direction to the center of the earth. A sun horizon scanner or limb sensor could be used to determine precisely the direction to the center of the solar disk. A moon horizon scanner may be used on lunar missions such as Apollo.

V. Horizon Scanning Accuracy

There are two basic limitations to the accuracy of any determination. One source of inaccuracy is errors developed in the instrument which makes the determination and the other source of error is variability in the physical phenomenon which is utilized for the determination.

A. Instrument Error

In the case of horizon scanners, the sources of instrument error are detector noise, mechanical tolerances, and mounting rigidity. The error caused by detector noise varies depending upon the type of pattern employed. Conical scan is more susceptible to detector noise than is linear scan. Detector noise can be made to have as small a contribution to the error of the attitude determination as desired by using larger optics, more sophisticated scanning methods, and better detectors. Mechanical tolerances in scanners can be reduced to better than a few seconds of arc, if necessary. In other words, the error in determining the local vertical can be made smaller than a few seconds of arc owing to mechanical tolerances of the scanner. Mounting rigidity can be a problem in some space vehicles. Errors can be developed as high as 1 deg if care is not taken. However, if it is desired to align an experiment with a heavenly body, alignment can be as good as 1/10 of a second of arc if the radiation coming into the experiment is shared with the attitude sensor.

B. Variability in Physical Phenomena

The other phenomena which cause errors in detecting local vertical are those due to the heavenly body in question. Errors which are dependent on the physics of the body generally are a certain magnitude of the apparent diameter of the body and decrease in absolute magnitude as the scanner gets farther away from the body because of the apparent decrease in absolute angular diameter. The error due to the ellipticity or oblateness of the surface of the earth could be made zero if one could take the ellipticity of the earth into account when calculating the attitude. Much larger errors can arise owing to irregularities in the apparent surface of the earth, such as might be caused by high clouds. These may be as large as 1/4 of a degree at an altitude of 200 nm but may be mitigated somewhat by the refraction by the atmosphere of the radiation from the tops of the clouds.

Another source of error for some types of scan is cold clouds appearing on the face of the earth. Figure 5 shows recordings from the detector in the horizon scanner in one of the Mercury flights, MA-5 [1]. In the left scan illustrated, the field of view of the sensor passed over a very cold cloud on the face of the earth. The temperature of the cloud appeared to be

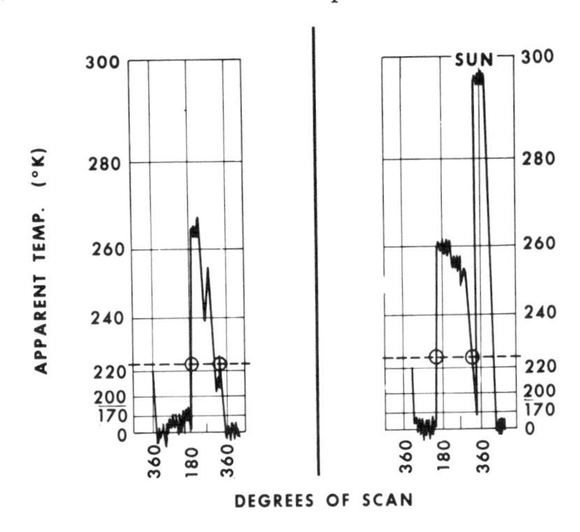

Fig. 5. Samples of horizon sensor signals from Mercury scanner. Spectral range, 2-15 microns.

about 205°K. Clouds colder than 200°K were observed during the flight. This type of cloud emits only about 1/2 of the total radiation emitted by an average area of the earth and only 1/4 of the radiation with wavelengths between 8 and 12μ that an average area of the earth radiates. These clouds could cause a great error in the attitude determined by any scanner whose field of view crossed them. The scanner might confuse the cloud with the horizon of the earth. The linear scan would not be bothered by this type of cloud appearing on the face of the earth. It would be bothered by such a cloud appearing at the horizon but the cloud would have a smaller apparent diameter and its radiation would be augmented by the atmosphere through which it would be viewed.

Dr. Rudolph Hanel of the Goddard Space Flight Center, Aeronomy and Meteorology Division, has suggested that by using a scanner sensitive only to wavelengths around 15μ, which are strongly absorbed by carbon dioxide, the phenomena of cold clouds could be avoided. Radiations in these frequencies should come from very high in the atmosphere at the top of the carbon dioxide region, above the level of cold clouds [2, 3].

The trace on the right in Fig. 5 was taken at a different time, when the sensor field of view passed near the sun. The signal from the sun appears to the right on the picture. The amplitude of the signal from the sun is clearly stronger than that from the earth, but the scanner could mistakenly confuse the onset of the signal from the sun with that from the horizon of the earth; an attitude determination from this data could be in error by many degrees. It is interesting to note from this data that the apparent diameter of the sun is about 73 deg which is much larger, of course, than the apparent diameter of the visible sun in the sky which is 1/2 deg.

The Tiros II, a weather satellite, carried a five-channel radiometer which sensed radiation from the earth in five different spectral bands. Figure 6 shows an example of the signals from the five-channel radiometer. The upper signal is from the radiometer which was sensitive to radiations with $6\frac{1}{2}\mu$ wavelength. This type of radiation is strongly absorbed by water vapor so the signal in this channel is from the top of the water vapor in the atmosphere. The signal from this channel shows severe limb darkening. In other words, the wave's shape from the sensor as it scans over the earth appears sometimes almost triangular rather than square and there is a smaller gradient of radiation at the horizon. The second line shows the signal from the radiometer which is sensitive to radiation of wavelengths of 8 to 12μ. This is considered to be the water vapor window and the radiation recorded here probably came from near the surface or from extremely opaque clouds. Figure 1 is an enlargement of the signal from this channel. Channels 3 and 5 were chiefly sensitive to visible light and were used as a check on the television camera in the satellite. The radiometer whose output here is channel 4 was sensitive to radiation wavelengths from 8 to 30μ. The signal in this channel is very similar to that of channel 2 whose radiometer was sensitive to 8 to 12μ radiation. Figure 8 shows another enlarged view of a signal from channel 2. The signals from the Tiros five-channel radiometer show the cold clouds and sharp gradients on the surface of the earth which also appear in the signals from the Mercury horizon scanner shown earlier [4].

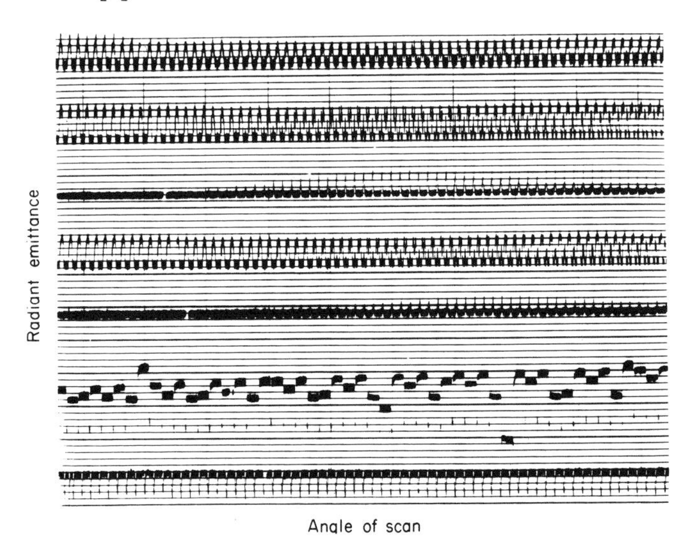

FIG. 6. Signals from Tiros radiometers.

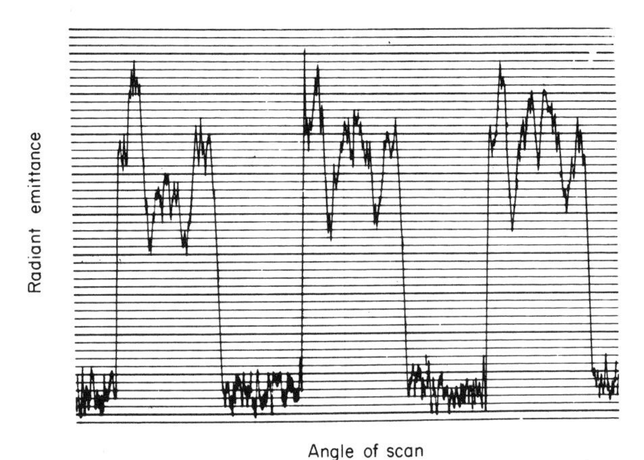

FIG. 7. Scans from Tiros radiometer. Spectral range, 8-12 microns.

The angular width of the signal from the earth was measured for 34 consecutive scans of the channel 2 radiometer. A signal level of about one-fifth of the maximum signal level was chosen as defining the horizon. The diameter of the earth appeared to vary from scan to scan. The rms variation was 1.8 angular degrees. This variation can be completely attributed to system noise. This variation in apparent diameter would cause errors of this approximate magnitude in attitudes determined from this data.

Mars is a planet which would look much like the earth to an infrared horizon scanner. It probably has few water clouds but it may have high, opaque dust clouds. In fact, one interesting, and as yet unexplained, fact about Mars is that its ellipticity, or oblateness, as calculated from the orbits of its planets is one-half that which can be measured on photographs of the planet. A reasonable hypothesis which explains this anomaly is that the solid surface conforms to the ellipticity predicted by the motion of the satellites, and the apparent bulge around the equator is caused by dust clouds which appear there. The actual ellipticity of Mars is somewhat less than that of the earth, but the apparent visible ellipticity of Mars is almost twice that of the earth.

This, or some other similar phenomena could also occur on earth, causing errors in horizon scanners. The apparent shape of the earth when viewed with radiation of certain wavelengths and the consistency of that shape with time has not been determined experimentally for the earth. This information is needed to predict the basic limit of accuracy of horizon scanners.

In the past, the attitude of spacecraft with respect to the sun has been determined by sensors which use the illumination from the total solar disk. However, sun spots can have an area equal to 1/300 of the total disk of the sun. A sensor which compares radiation from the two halves of the solar disk could be thrown off by 5 seconds of arc. For a really precise determination of the direction of the sun, some sort of limb tracking or horizon sensing of the sun must be used.

C. Experimentally Determined Accuracy

The Tiros satellite also carried another infrared-sensitive scanner, called the horizon pipper, besides the five-channel radiometer. The attitude of the satellite was calculated by reducing the horizon pipper data and also by using information from the television pictures taken by the satellite. The positions of the horizons and other landmarks appearing on the pictures were used to determine the satellite's attitude at the time they were taken.

By hand selecting the best data from the scanner and smoothing it with a computer, the attitude determined using the horizon scanner data corresponded to that determined from the television pictures to within 1 or 2 angular degrees. The attitude determined from the television pictures is considered to be more accurate. So the inaccuracy in the attitudes determined by the horizon scanner was 1 or 2 deg.

The horizon scanner on the Mercury capsule was specified to be accurate to 1 deg of arc. Figure 5, recorded from the Mercury scanner, indicates that the scanner at certain times failed to remain within the 1 deg error specified.

VI. Conclusion

Horizon scanners for attitude determination, up to now, have been limited by instrument accuracy or design rather than by physical phenomena intrinsic to the body being scanned. The accuracy of scanners which have been flown has been, at best, 1 or 2 deg. It appears that accuracy of 2/10 of a degree for altitudes below 1000 miles and 1/10 of a degree for altitudes above 1000 miles are attainable, using techniques which are now in development.

The use of physical phenomena other than the total infrared radiation from the earth, such as air glow or narrow spectral band infrared radiation, may be necessary to attain better accuracy. More must be known about these phenomena before horizon scanners can be designed to utilize them.

References

1. Report on Analysis and Correlation of Data from MA-5 Mission, *Barnes Engineering Co. Rept.* 4858-*SR* 1, August 20, 1962.
2. R. A. Hanel, W. R. Bandeen, and B. J. Conrath, The infrared horizon of the planet earth. *J. Atmospheric Sci.* **20**, 73-86 (1963).
3. J. W. Burn, The application of the spectral and spatial characteristics of the earth's infrared horizon to horizon scanners. *I.E.E.E. Trans. Aerospace-Support Conf. Procedures*, pp. 1115-1126, August 4-9, 1963.
4. T. L. Altshuler, Infrared Transmission and Background Radiation by Clear Atmospheres, *General Electric Co. R* 61 *SD* 199 (December, 1961).

Earth Scan Analog Signal Relationships in the Tiros Radiation Experiment and Their Application to the Problem of Horizon Sensing

BARNEY J. CONRATH

Goddard Space Flight Center, National Aeronautics and Space Administration, Greenbelt, Maryland

I. INTRODUCTION

SEVERAL OF A SERIES of Tiros meteorological satellites have been placed into near-circular orbits by NASA, and more are planned for the future. Tiros I, launched on April 1, 1960, carried cloud-cover television cameras. Subsequent Tiros satellites have contained, in addition to television cameras, instrumentation for measuring radiation emitted and reflected by the earth and its atmosphere. The system aspect of the radiation experiment has been described by Bandeen *et al.* [1].

One of the several types of satellite-borne radiometers employed [2, 3], is a five-channel medium-resolution scanning radiometer. This instrument's primary purpose is to obtain measurements applicable to problems in meteorology and atmospheric physics, such as albedo determination, heat balance, and cloud-cover determinations at night or when other means of observation are not available. Some preliminary results obtained from these measurements have been published [4, 5]. In addition to the primary objectives of the experiment, there is considerable interest in applying the simultaneous measurements made in several spectral regions to problems encountered in the design of horizon sensor systems.

Two basic problems in selecting the best spectral region for a horizon sensor are the effects of cloud systems on the sensor's behavior and its dependence on the viewing angle (limb darkening), especially at large angles. Several examples which are applicable to these problems have been chosen from the Tiros data and are presented in this paper to illustrate the typical behavior of the various channels under different synoptic situations.

II. Instrumentation and Calibration

The scanning radiometer consists of a cluster of five sensors with coincident fields of view about 5 deg wide at the half-power points of the response. The spectral regions covered by these five channels are

(1) 6.0 to 6.5 μ, the water vapor absorption band;
(2) 8.0 to 12 μ, the atmospheric "window";
(3) 0.25 to 6.0 μ, the total reflected solar radiation;
(4) 7.0 to 30 μ, thermal radiation; and
(5) 0.55 to 0.75 μ, the visible reference.

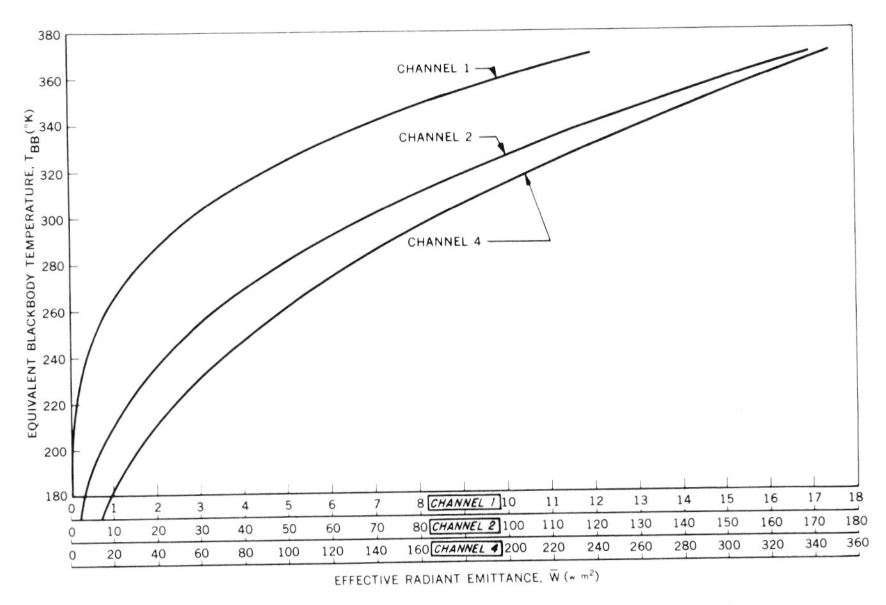

Fig. 1. The effective radiant emittance versus the equivalent blackbody temperature for the three thermal channels used on Tiros III. The curves are obtained by integrating the product of the Planck function and the spectral response curves over all wavelengths.

These figures are purely nominal. The effective spectral response curves of each channel are given in references [1, 6].

The optical axis of the instrument points in two directions 180 deg apart. By means of rotating choppers, the sensors are exposed in rapid succession to radiation from first one direction and then the other. The resulting output signal is proportional to the difference in flux received from the two directions. The optical axis is inclined 45 deg to the satellite's spin axis; thus the radiometer scans the earth as the satellite rotates about its spin axis, and the orbital motion of the satellite provides the scan advance. Whenever one direction of the optical axis points toward the earth, the other direction is pointed toward outer space. Therefore, outer space can be used as a zero reference. The instrumentation is described completely in references [1, 7].

The three thermal channels (1, 2 and 4) are calibrated by exposing the instrument to a blackbody source at various known temperatures. From these temperatures, the portion of the radiant emittance which falls within the spectral response of a given channel, $\bar{W}$, is computed by integrating the Planck function over the spectral response curve (Fig. 1). Thus, the outputs of the thermal channels can be expressed in terms of either $\bar{W}$ or the equivalent blackbody temperature. The visible channels are calibrated in a similar fashion by exposing the instrument to a target of known spectral radiant emittance. The calibration process is described in detail in reference [6].

III. Selected Examples

The following cases have been selected in an effort to provide examples which will be of interest in horizon sensor considerations. Several of the examples are from previously published case studies [5]. In all cases, the data shown are from Tiros III.

Where possible, examples have been chosen for which television coverage exists. The pictures have been gridded in terms of geographic longitude and latitude, and the radiometer scan paths on the earth's surface have been drawn. Analog oscillograms of the radiometer output for these scan paths are included. Arbitrary angles of satellite rotation are indicated along the scans in the television pictures and along the abscissas in the oscillograms. Thus, points along a scan may be correlated with corresponding points on the oscillogram. The ordinates of the oscillograms are given in terms of effective radiant emittance $\bar{W}$.

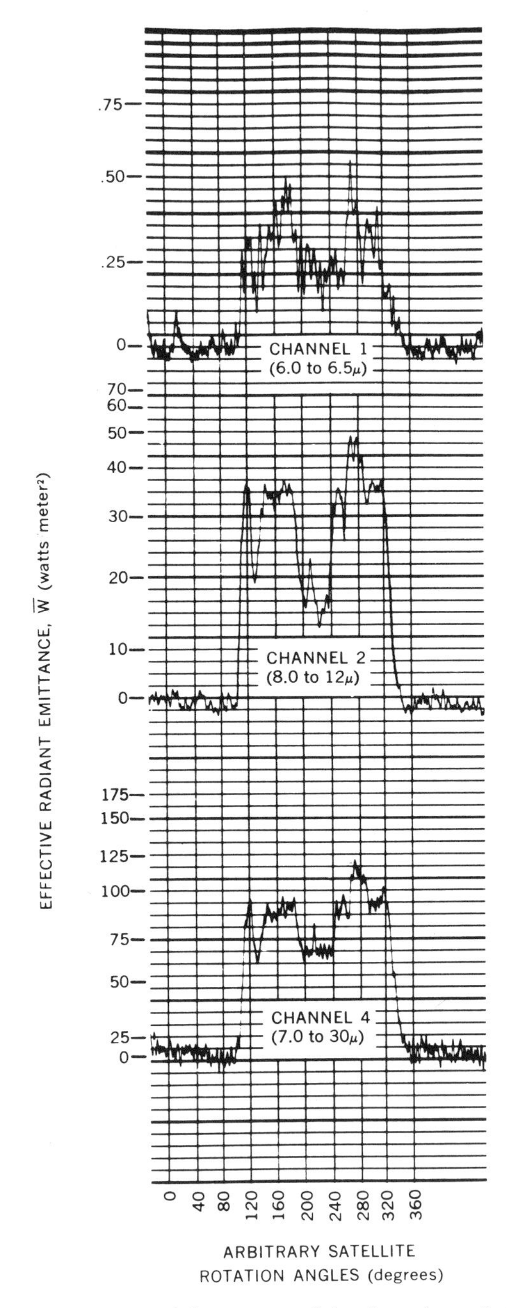

FIG. 2. Analog oscillogram of the outputs of the three thermal channels showing the data discussed in Example 1.

Example 1

The first example is taken from one of three case studies previously described in reference [5]. Figure 2 shows the oscillogram of the three thermal channels corresponding to the scan path indicated on the photograph in Fig. 3. A large cloud mass is prominent near the center of the picture, with a clear area over the Great Lakes and the Michigan peninsula in the upper left corner.

The situation illustrates the behavior of the sensors when scanning an extensive cloudy region of fairly high albedo. This was the largest

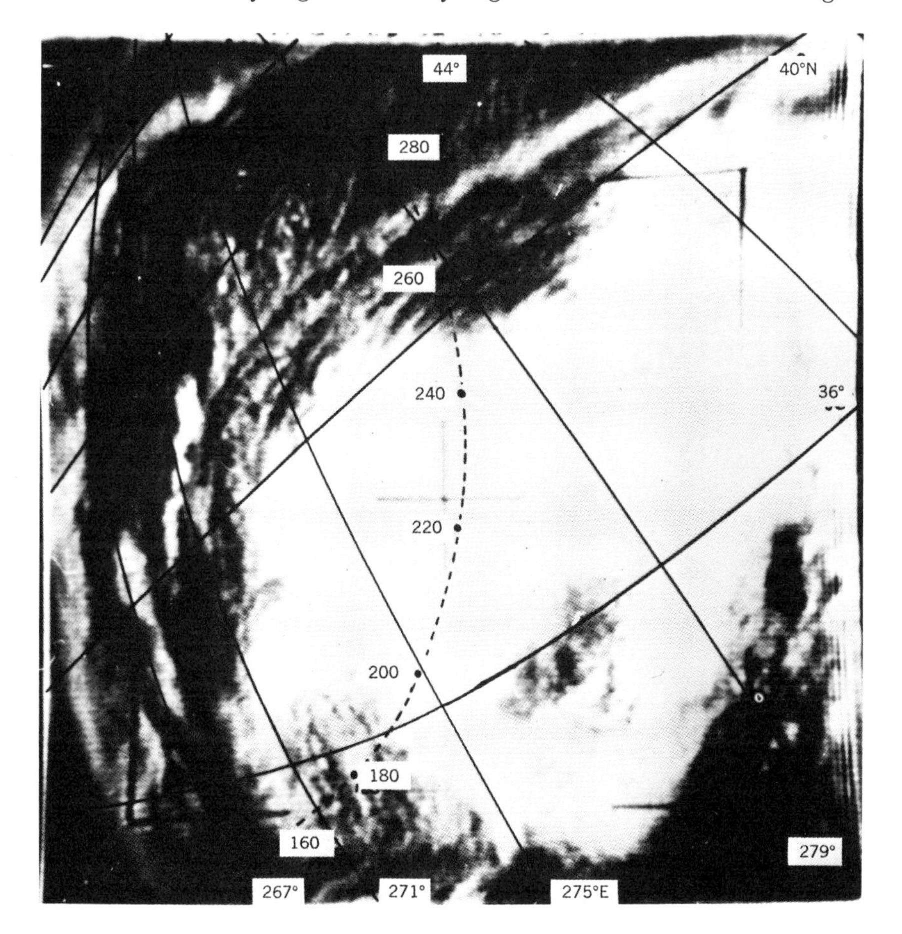

FIG. 3. Photograph taken by Tiros III at about 1735 GMT on July 12, 1961. Michigan and the Great Lakes are visible in the upper left corner. The radiometer scan path (dashed), with arbitrary satellite rotation angles, is shown crossing the large central cloud mass.

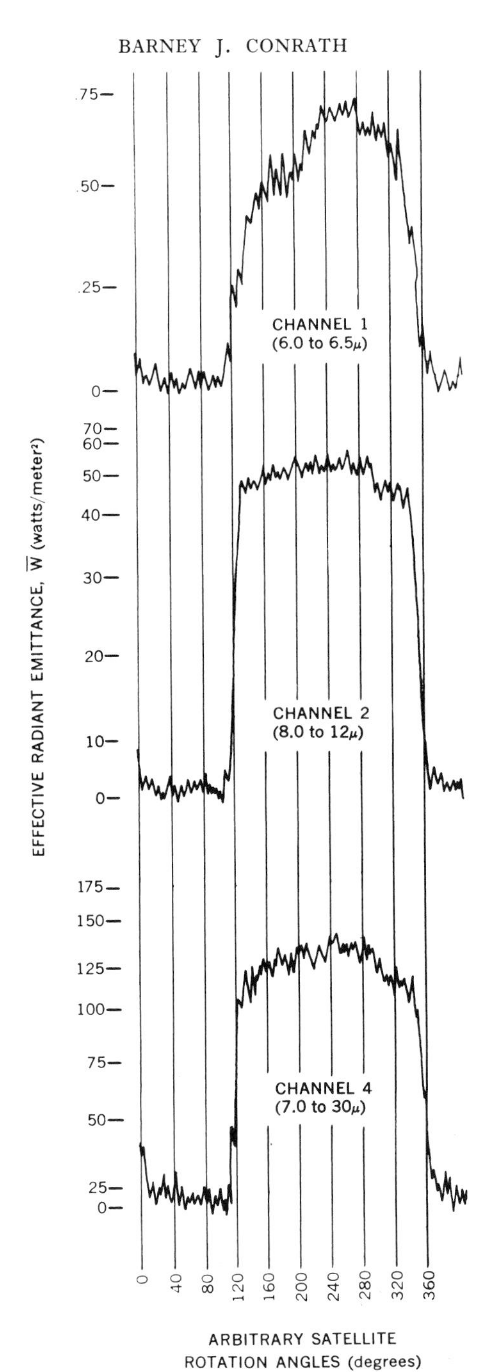

FIG. 4. Analog oscillogram showing the data from the three thermal channels discussed in Example 2.

area of solid cloud cover that could be found in a search of the pictures from the first 200 orbits of Tiros III. The maximum albedo measured in this area was about 55% in the 0.55 to 0.75 μ region.

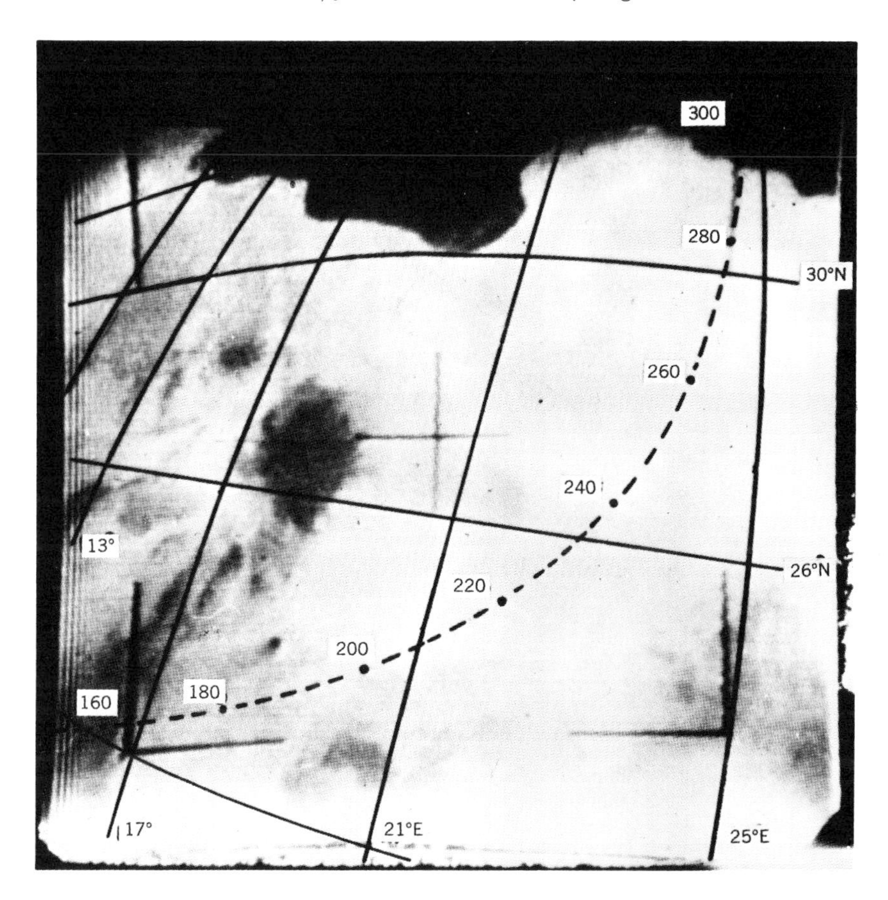

FIG. 5. Photograph taken over Libya by Tiros III at about 1053 GMT on July 15, 1961. The North African coast line and the Mediterranean See are visible near the top of the picture.

The most obvious feature of this example is the large decrease in radiant emittance seen in all three thermal channels as the cloud mass is scanned. The radiation gradients viewed through the Tiros 5-deg fields of view are quite steep in the 7 to 30 μ channel and especially in the 8 to 12 μ channel where they are equal in magnitude to the horizon gradients. The gradients in the 6.0 to 6.5 μ channel are somewhat less steep. It is not possible to show the entire area covered by a radiometer scan path in a single television picture; hence, conditions near the points where the scan path intersects the horizon cannot be seen here. However, it is probable that the horizon was clear at the points of intersection since no strong signals were observed in the visible channels and the weather map shows no overcast in these regions.

Example 2

The second example, also taken from reference [5], is concerned with data acquired over the Libyan desert. The oscillogram is shown in Fig. 4 and the television picture in Fig. 5. The brighter areas are sand covered for the most part, having albedos of 27 or 28 % as measured by the 0.55 to 0.75 μ channel. This region is apparently totally free from cloudiness.

The oscillogram (Fig. 4) of the scan path, especially between 200 and 280 deg of satellite rotation, illustrates the behavior of the thermal channels over a region of what must be fairly uniform radiant emittance. In this region, the effective blackbody temperatures deduced from the 8 to 12 μ channel are around 310°K, which is near the channel's saturation point. The high response observed in the 6.0 to 6.5 μ channel indicates the absence of any large amount of water vapor in this area.

At angles of satellite rotation less than 200 deg, the radiometer was viewing the uplands region near the border of Libya and Chad. The most outstanding feature of this region is the lower radiant emittance values viewed by channel 1, which indicate a higher atmospheric water vapor content. Near 295 deg of satellite rotation, the transition from land to water shows up sharply in channels 2 and 4.

Example 3

The following example, also from reference [5], was chosen to illustrate the behavior of the sensors over a relatively clear tropical ocean. The data were taken by Tiros III off the northern coast of South America. The oscillogram is shown in Fig. 6, and the television picture in Fig. 7. The coast line of the Guianas, partially obscured by clouds, can be seen in the upper left-hand corner of Fig. 7.

The albedo measured by channels 3 and 5 is quite low over this area, as would be expected. However, the effective blackbody temperature obtained from the 8 to 12 μ channel is about 20°K lower than the expected surface temperature of the water in this region. This is probably caused, in part, by residual water vapor absorption in the atmospheric "window," since soundings at Trinidad indicated over 5 cm of precipitable water

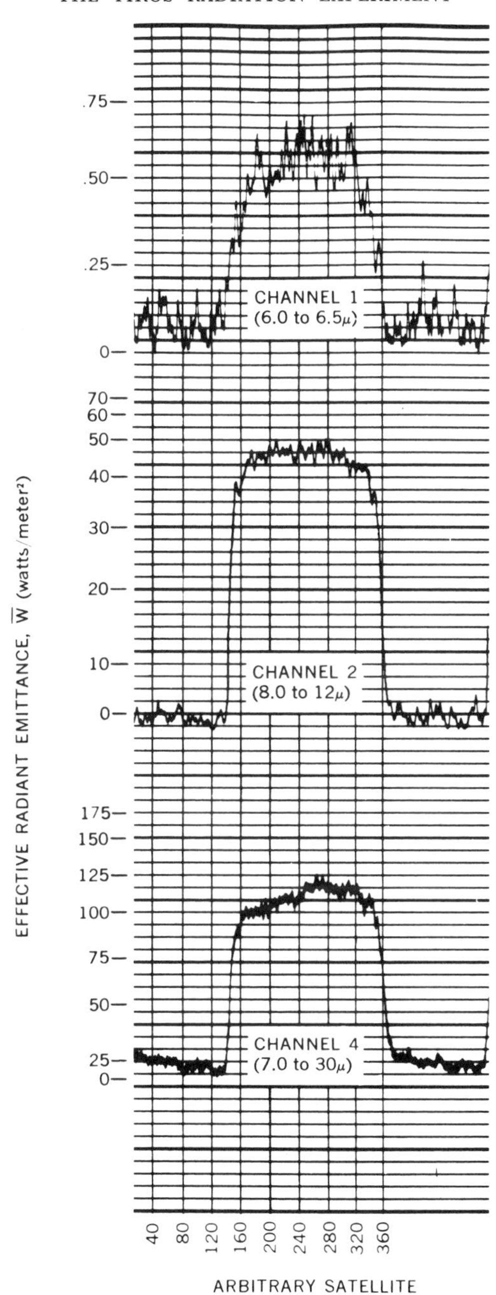

FIG. 6. Analog oscillogram showing the data from the three thermal channels discussed in Example 3.

vapor. Other contributing factors may be the presence of high cirrus clouds (which do not show in the television picture) or the presence of other absorbers, such as aerosols.

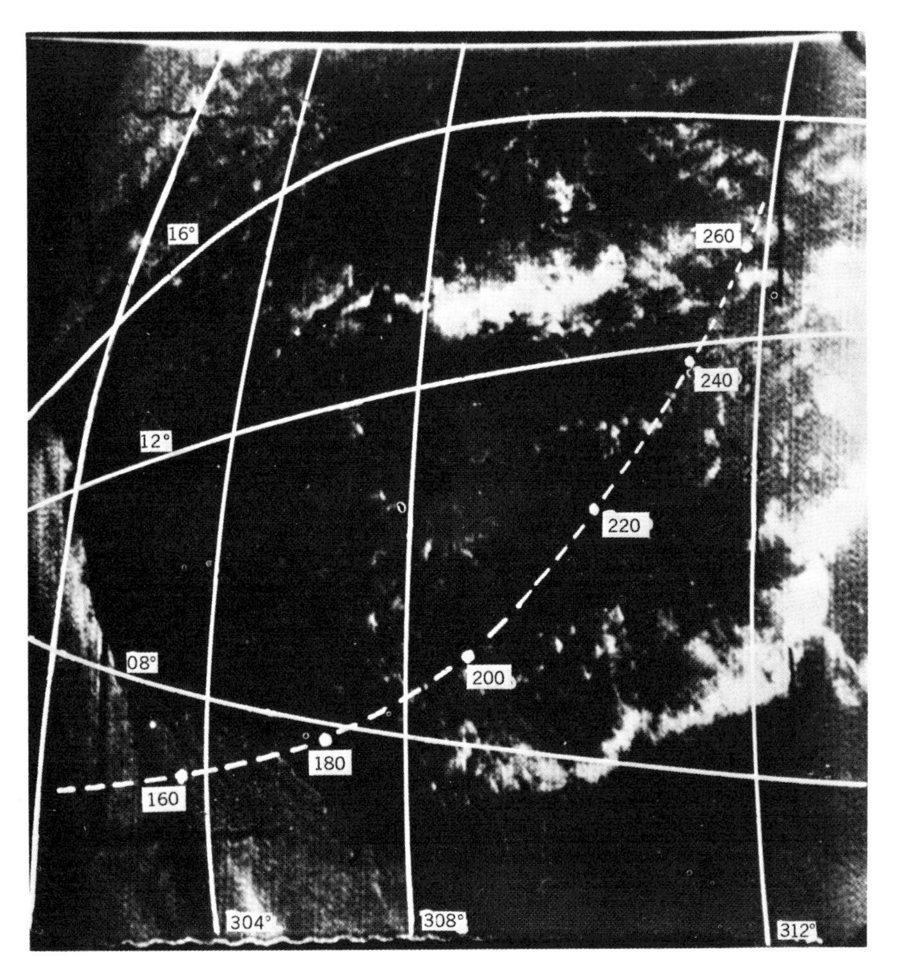

FIG. 7. Photograph taken off the northern coast of South America by TIROS III at about 1444 GMT on July 20, 1961. The partially cloud-covered coast line of the Guianas is seen in the lower left corner.

If it is assumed that the area covered by this scan path is fairly uniform in radiant emittance as the picture would seem to indicate, this case provides a good example of the variation of the sensor output with viewing angle. At the center of the scan, the zenith angle (at the spot viewed) is near zero. This angle increases in either direction from the center of the scan, approaching 90 deg at each horizon. The apparent limb darkening is least for the 8 to 12 μ channel and greatest for the 6.0 to 6.5 μ channel.

Example 4

The next case illustrates the behavior of the sensors as a cloudy horizon is scanned. The oscillograms of the visible channels (3 and 5) have been included in place of television pictures since it is difficult to determine the conditions near the horizon from the pictures with any certainty.

This scan path begins at a point off the eastern coast of Florida and terminates in Canada about 50 deg N and 70 deg W north of Maine. The visible channels (Fig. 8) show relatively strong signals in the latter region (between 280 and 320 deg of satellite rotation), which would indicate the presence of clouds on this horizon. In addition, data from the northern hemisphere surface chart (Fig. 9) indicate a low pressure area and stationary front with associated cloudiness in this region. In the thermal channel oscillograms (Fig. 10), the 8 to 12 μ channel shows the largest relative decrease in amplitude. The 7 to 30 μ channel shows less decrease, and the effect is almost lost in the 6.3 μ channel owing to the large dependence of this channel on the viewing angle. The remainder of the scan between 140 and 280 deg of satellite rotation is over clear ocean with occasional broken cloudiness.

Example 5

The next example concerns data acquired on July 17, 1961, at about 1308 GMT over northwest Africa. The analog oscillograms for the thermal channels are shown in Fig. 11, and the picture of the region, containing part of the African coast line and the Atlantic Ocean, in Fig. 12. That portion of the cloud system intersected by the scan path displayed the highest albedo found in a search of all Tiros III radiation data for which corresponding television pictures existed. The maximum measured albedo was 62% in the 0.55 to 0.75 μ region.

The oscillograms show a narrow but very deep minimum at a rotation angle of about 220 deg, which corresponds to the region of maximum observed albedo. The thermal channels show a reversal of their normal pattern in that channel 1 shows the highest equivalent blackbody temperature, followed by channel 4, with channel 2 the lowest. This indicates a very high cloud, probably near the tropopause. In this example, the profile of the channel 1 analog more closely follows those

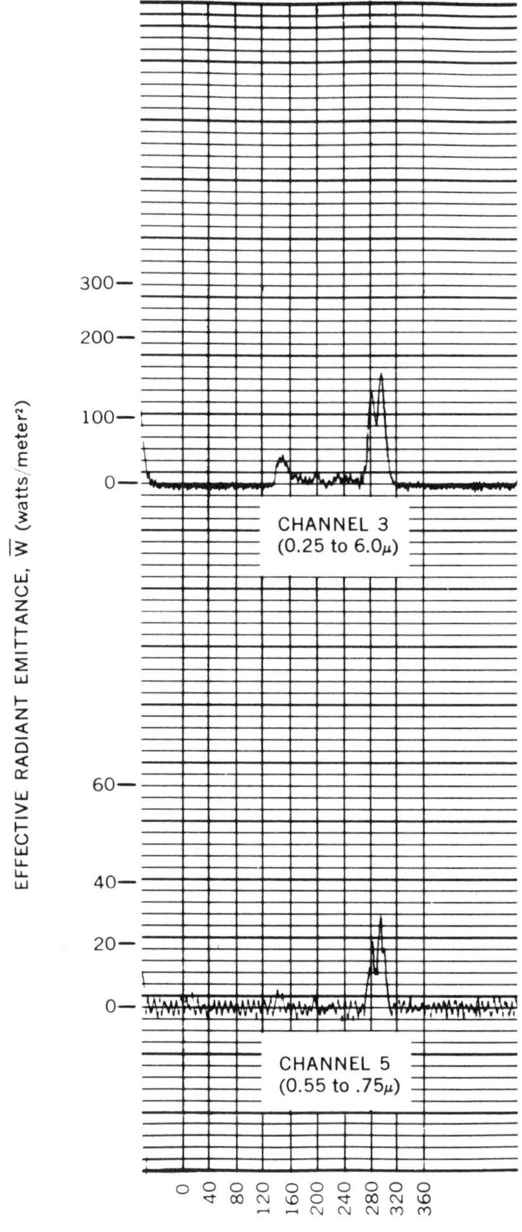

FIG. 8. Analog oscillogram showing the data from the two visible channels discussed in Example 4.

of the other two channels than in the previous cases. Channel 2 again shows the sharpest gradients, while those of channel 4 are less acute.

For rotation angles of less than 200 deg, the pattern is typical for a region of broken cloudiness. At angles greater than 260 deg, the sensors are probably viewing a vegetated area that is relatively cloud-free.

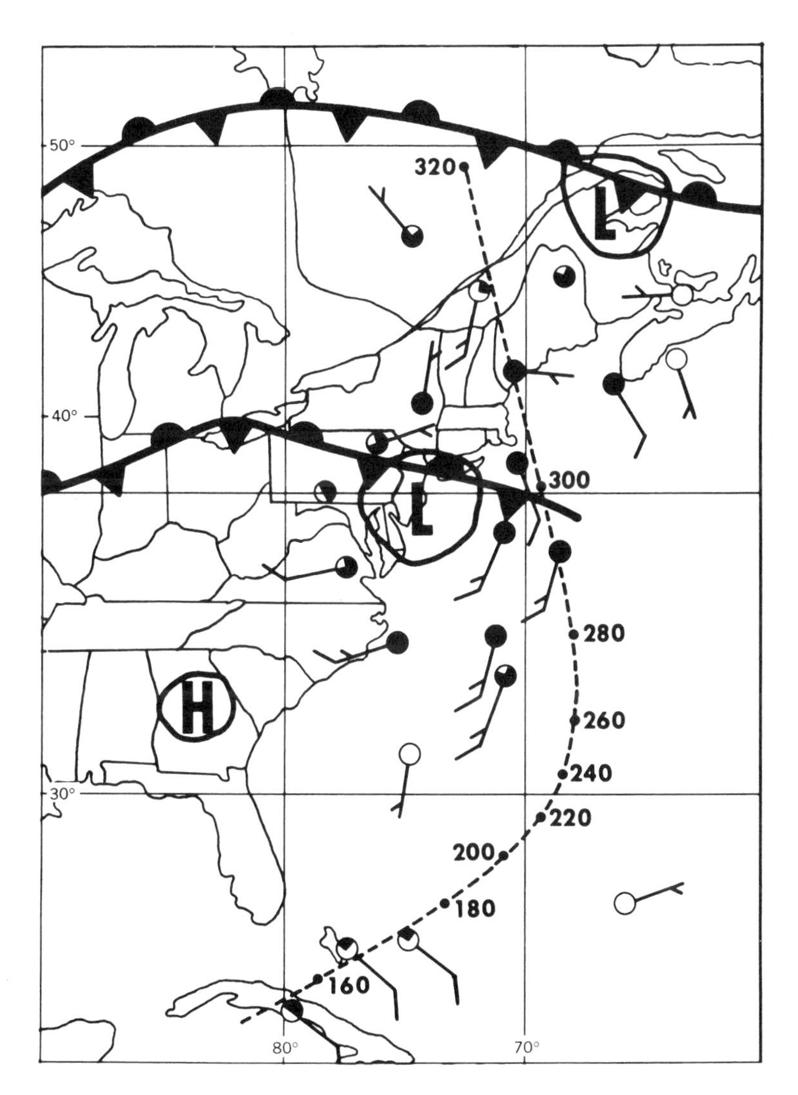

FIG. 9. Map showing data taken from the northern hemisphere surface chart for 1800 GMT on July 17, 1961. The broken line shows the satellite scan path discussed in Example 4.

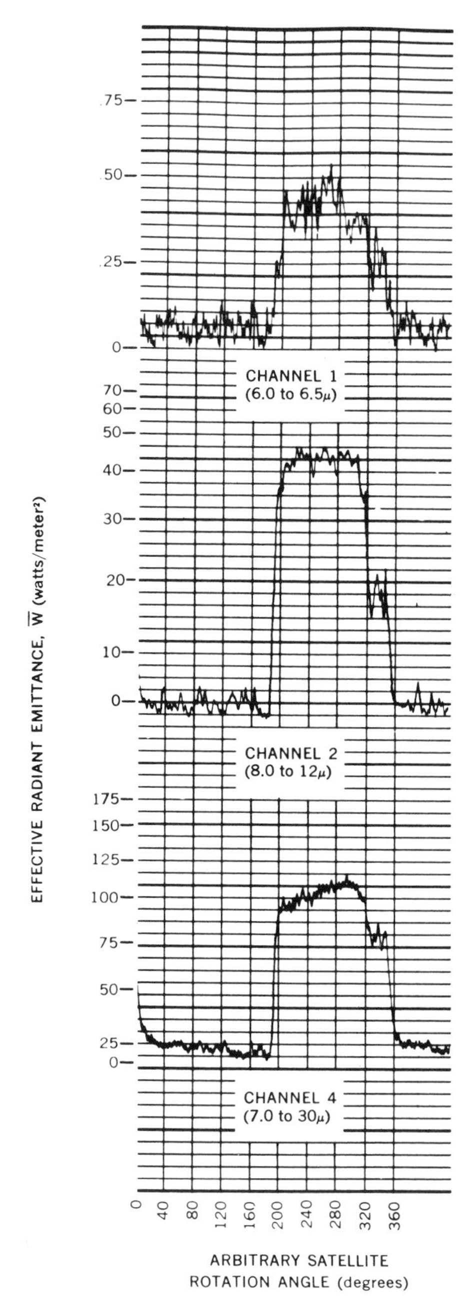

FIG. 10. Analog oscillogram showing the data from the three thermal channels discussed in Example 4.

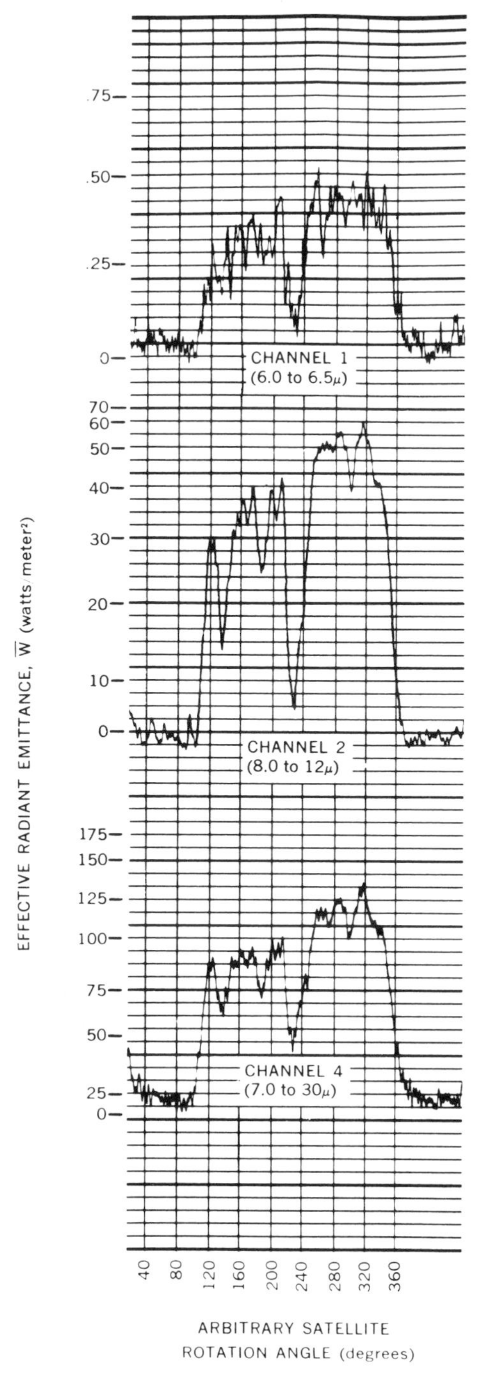

FIG. 11. Analog oscillogram showing the thermal data discussed in Example 5.

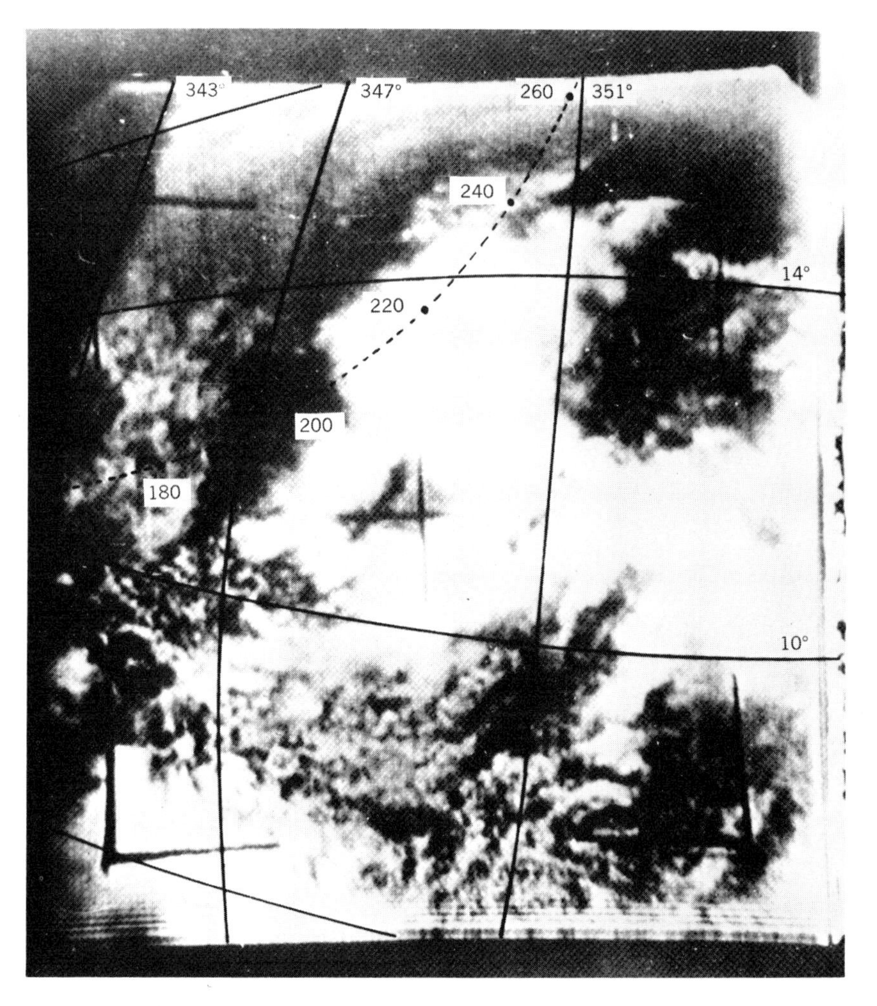

FIG. 12. Photograph taken over northwest Africa by Tiros III at about 1309 GMT on July 17, 1961. The African coast line and the Atlantic Ocean can be seen in the upper left corner.

Example 6

The data presented in this final example were taken by Tiros III over the Caribbean on July 22, 1961, at about 1512 GMT. Figure 13 shows the analog oscillograms of the three thermal channel outputs as the radiometer scanned the path seen in Fig. 14. The cloud system associated with Hurricane Anna occupies the central portion of the picture; Florida and the Gulf of Mexico lie to the upper left.

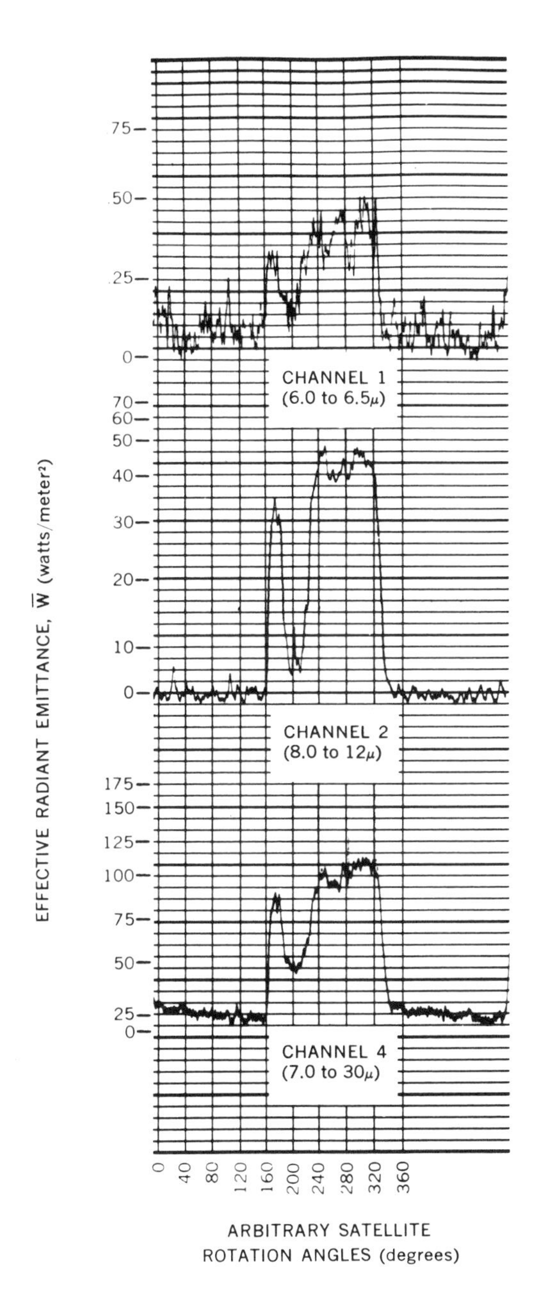

FIG. 13. Analog oscillogram of the thermal data discussed in Example 6.

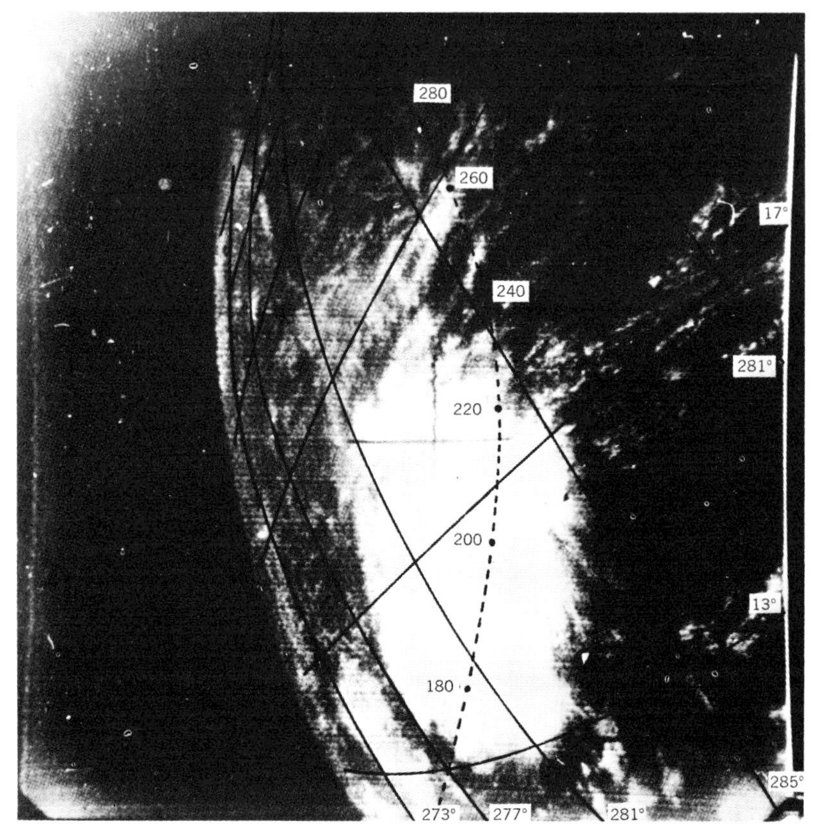

FIG. 14. Photograph of Hurricane Anna taken by Tiros III at about 1512 GMT on July 22, 1961. Cuba and Florida can be seen to the north of the storm center.

IV. SUMMARY AND CONCLUSIONS

It would be possible to choose examples illustrating sensor behavior under many conditions, but the above cases cover a sufficiently wide range to provide some feeling for the characteristics of the various spectral regions. Examples 1, 2, 3, and 4 may be considered typical of the situations described. The extent of the cloud system in Example 1 is probably somewhat greater than might normally be expected, but the behavior of the sensors can be considered typical. Although the synoptic situations illustrated in Examples 5 and 6 are relatively rare, they show the extreme effects that isolated storm centers can produce in the infrared output.

The 6.0 to 6.5 μ region has a serious disadvantage in that it usually displays severe limb darkening. This is noticeable in all of the examples, but is best illustrated in Example 3 (Fig. 6). This effect is expected to be greatest in the tropical atmosphere and least in the arctic atmosphere [8]. Another disadvantage of this spectral region is the small amount of total radiant power available, which leads to instrumentation difficulties in obtaining a good signal-to-noise ratio.

The 8 to 12 μ channel shows an excellent reponse to the horizon with steep gradients and little limb darkening. In addition, there is no signal-to-noise problem because of the large amount of total power available. However, the pronounced reaction to the presence of clouds almost certainly makes this spectral region unsuitable for use in horizon sensor work, as Examples 1, 4, 5, and 6 well illustrate. The gradients displayed as the sensor scans the cloud systems are as steep as those resulting from the horizon itself, and in the case of very high clouds, as in Examples 5 and 6, the signal can plunge almost to the noise level.

The 7 to 30 μ channel is affected by limb darkening only slightly more than the 8 to 12 μ channel, and ample total radiant power is available. Since the atmospheric "window" is also included in this spectral region, the signal is considerably affected by cloud systems. However, this effect is somewhat more moderate here than in the 8 to 12 μ region as Examples 1, 5, and 6 demonstrate. In general, although the gradients are not as steep and the drops in radiation level when scanning clouds are not as severe, these effects will continue to cause serious problems.

The Tiros data indicate that the final choice of an optimum spectral region should be one which does not include the atmospheric "window". In fact, it would seem more profitable to investigate spectral regions not directly covered in the Tiros experiments, such as the 15 μ carbon dioxide band.

References

1. W. R. Bandeen, R. A. Hanel, J. Licht, R. A. Stampfl, and W. G. Stroud, Infrared and reflected solar radiation measurements from the Tiros II meteorological satellite. *J. Geophys. Res.* **66**, 3169-3185 (1961); also *NASA Tech. Note D-1069*, November 1961.
2. R. A. Hanel, Low resolution unchopped radiometer for satellites. *ARS J.* **31**, 246-250 (1961); also *NASA Tech. Note D*-485, February 1961.
3. R. W. Astheimer, R. De Waard, and E. A. Jackson, Infrared radiometric instruments on TIROS II. *J. Optical Soc. Am.* **51**, 1386-1393 (1961).
4. "TIROS II Radiation Data Catalog," Vol. I. NASA, Goddard Space Flight Center, August 15, 1961.
5. W. Nordberg, W. R. Bandeen, B. J. Conrath, V. Kunde, and I. Persano, Preliminary results of radiation measurements from the TIROS III meteorological satellite. *J. Atmos. Sci.* **19**, 20-30 (1962); also *NASA Tech. Note D-1338*, in publication, 1962.

6. "TIROS II Radiation Data Users' Manual." NASA, Goddard Space Flight Center, August 15, 1961.
7. J. Davis, R. A. Hanel, R. A. Stampfl, M. Strange, and M. Townsend, Telemetering IR data from the TIROS II meteorological satellite. *6th Nat. Symp. Space Electronics and Telemetry, Albuquerque, New Mexico, September, 1961*; also *NASA Tech. Note D-1293*, in publication, 1962.
8. R. A. Hanel, and D. Q. Wark, TIROS II radiation experiment and its physical significance. *J. Optical Soc. Am.* **51**, 1394-1399 (1961); also *NASA Tech. Note D-701*, December 1961.

Author Index

Numbers in parentheses are reference numbers and are included to assist in locating references when the authors' names are not mentioned in the text. Numbers in italics refer to the page on which the reference is listed.

Subject Index